PRAISE FOR
NEITHER AN OFFICER NOR A GENTLEMAN

"Lewis Smith provides an excellent addition to all that has been written about the Vietnam War. He gives a realistic portrayal of life on the home front and of the air crews that served on aircraft carriers while supporting ground operations in close combat. His book is simply a first-class reading experience."
— Burton D. Patrick,
Lieutenant General, U S Army (Retired)

"Lewis Smith's well-written, intimate, firsthand account of life on an aircraft carrier during the Vietnam War is a walk down memory lane. It is a rare view into an enlisted man's life that will keep the reader enthralled. His war stories are what everyone who has never experienced war needs to hear. Not many decorated veterans recount their stories to the uninitiated, and certainly not in such an entertaining manner. The author gives the reader a unique view and understanding of a young man's great adventure and the Southern history of the late 60s that helped form it."
— Mayor Wiley Johnson, Summerville, SC,
LtCol, USAFR, Vietnam era to Desert Storm

"Lewis Smith's latest book is a coming of age story of a young enlisted man serving on an aircraft carrier during the Vietnam War years. Sort of Catcher in the Rye meets Naval Aviation, it is the story of his time in the Navy. It covers the good and the bad, and offers insight into the civil rights era and the issues the Navy had with drugs during that period. The author seems to have a knack for ending his chapters with a pithy, usually humorous comment that is like an extra reward at the chapter end. Brutally honest and sometimes cringe worthy, it's an interesting read that frankly depicts that time in our history."

—Gary W. Edwards,
Captain, US Navy (retired)

NEITHER AN OFFICER NOR A GENTLEMAN

NEITHER AN OFFICER NOR A GENTLEMAN

F. Lewis Smith, AQB2

Deeds Publishing | Athens

Copyright © 2020 — F. Lewis Smith

ALL RIGHTS RESERVED — No part of this book may be reproduced in any form or by any electronic or mechanical means, including information storage and retrieval systems, without permission in writing from the authors, except by a reviewer who may quote brief passages in a review.

Published by Deeds Publishing in Athens, GA
www.deedspublishing.com

Printed in The United States of America

Cover design by Mark Babcock.

ISBN 978-1-950794-20-1

Books are available in quantity for promotional or premium use. For information, email info@deedspublishing.com.

First Edition, 2020

10 9 8 7 6 5 4 3 2 1

This is dedicated to the one I love. I dedicate this book to the best proof reader, advisor and honest critic a man ever had: JoAnn, my beautiful wife of 50 years who has lovingly sat by my side researching and answering our questions each of the ten times she's read this memoir.

CONTENTS

Prologue	xi
1. Dr. Martin Luther King	1
2. Boot Camp To Millington	13
3. Marine Sergeant Acker	25
4. Foggy Windows	37
5. Virginia Beach	53
6. San Diego Softballs	67
7. Burial At Sea	83
8. Aircraft Carriers	97
9. Japanese Hot Tubs	109
10. Filipino Chicken-On-A-Stick	119
11. The Pigeon And Ghost Fronts	131
12. Burials And Bombs	143
13. Peanuts And Train Stations	157
14. Wedding Bells And Wives	171
15. The Funeral Guest	183
16. The Rescue Of The Flamingo	195
17. Chava Laska's Home Army	205
18. Pompeii And Calley	223
19. Corfu And My Lai	237
20. Crete And Collision At Sea	251
21. The Dead And The Quick	267
22. What Happened To Calley	279
Photographs	299
About the Author	313

PROLOGUE

Although I served honorably in the United States Navy for four years, one on Yankee Station in the Gulf of Tonkin on an aircraft carrier, I still feel funny calling myself a Vietnam veteran. But that's what I am; I've got the medals, the paperwork, and the combat pay to prove it. I just haven't bought the hat yet. I know being shot at in-country by the Viet Cong or NVA soldiers is not the same as being on a floating ammo dump loaded to the brim with fighter and attack jets, but many a Navy man died off the coast of Vietnam and in its coastal waters serving their country. More importantly, there were no greater heroes than the courageous Navy aviators and aircrewmen who daily risked their lives bombing enemy targets and supporting our ground troops. Those aviators were killed or taken prisoners on a daily basis at many remote places during the war. The men of all branches of American service who served during Vietnam and stared death right in its face are the real Vietnam vets, not me. Those feelings, if even misplaced, may be the reason I've never joined a military service group or have ever attended a ship or squadron reunion.

I've probably thought about the war every day since I left the service in 1971, but it was casual at best until five or six years ago. I began writing for fun and had written two books by the time I decided to put some of this on paper. During the years, I'd jotted down some stories, a few were published in my local newspaper, and then I took a couple of courses in writing non-fiction at a local university. I was finally ready to begin. I have read extensively (the best asset to good writing is a tremendous amount of good reading) and I hope it shows. I have used four period and pertinent Time, Newsweek and Esquire magazines from 1970 and 1971, three scholarly hardbacks specifically concerning events in this book and nine other significant books about the U.S. Navy and the war in Vietnam. I have constantly referred to my two cruise books from deployments to the western Pacific Ocean and the Mediterranean Sea. The first was the U.S.S. *Constellation* 1969-1970 cruise book, and the second was the U.S.S. *Forrestal* 1971 book. To supplement that material I have spent two months fact-checking my information by using legitimate internet sources and official websites. I have discussed the war with some folks and have just talked with many others. My wife dug out the letters I wrote her when she was in college or I was at sea and gave me permission to use parts of them here. She has read and reread this and earlier versions at least ten times as I fought through my material and rewrote everything.

My publisher, Lt. Bob Babcock, U.S. Army, was an Army infantry platoon leader in-country in '66 in Vietnam and has guided me as I moved forward. I received reviews and got constructive criticism from three friends: Lt. Gen. Burton D. Patrick, U.S. Army (ret), former Commanding General of the 101st Airborne Division; Captain Gary W. Edwards, U.S. Navy (ret), former Captain of the nuclear submarine U.S.S. *Tennessee* (SSBN-734); and Wiley Johnson, LtCol, USAFR (ret), Vietnam era to Desert Storm, a former Mayor of Summerville, SC.

This is a true story of my years in the Navy from 1967 through 1971 and their contemporary events, but I've added a couple of minor tales of no consequence, tales I heard but did not see, just for color. The dialogue, obviously, is a generalized version of what it actually was.

Neither an Officer nor a Gentleman

No memoir from 50 years ago could be expected to be exact, but this gives you the gist of it. My character Pete was a composite of primarily myself, my much older cousin Freeman Irby "Pete" Gibson and my best friend Gabriel Xavier Michelena, both now dead. Pete was a good ole boy from Georgia who loved to eat, drink, and party. One time I had to bail him out of jail in Thomson, Georgia, for pulling a flim-flam at Walmart. My best friend was a handsome Basque from Miami who was just a regular laid-back guy. This was written to be enjoyable; a candid, funny, irreverent book, just as I am. My friends and I cussed more often than this book shows and used worse words most of the time, but my wife made me tone it down so our Sunday school class could read it. I doubt I have accomplished that mission. But keep in mind, this is not a technical manual or a history book. It was meant to be enjoyable, and it may be, since I was neither an officer nor a gentleman.

—F. Lewis Smith
Thomson, Georgia
April 8, 2020

1. DR. MARTIN LUTHER KING

At six-thirty in the evening of April 4, 1968, all sailors were called back from the chow hall at Naval Air Station Millington and put under lockdown in our barracks. Everything on the base was shut down, and we were worried; we had no idea what was happening. Our avionics instructor Sergeant Acker knew but refused to say a word. Someone tried to blow up the base, one rumor said, and a sailor was killed. As we were corralled into our barracks, the black sailors in the school were taken aside and led away. Once they had all been gathered into the base gymnasium by themselves, huddling close like a flock of black sheep, wondering what was happening to them this time, the First Class Petty Officer in charge of our barracks called us together. It was seven-thirty, and we were finally told that Dr. Martin Luther King, Jr. had been shot twenty-three miles away in downtown Memphis.

"Listen up, guys. Sorry about the confusion and the long wait, but I've got some bad news. Martin Luther King was shot a couple of hours ago in downtown Memphis. I know it may not be that earth-shattering to you, but our black sailors are really going to be tore up about it. They're going to try to burn this place down and kill every redneck they see. They will lash out at every cruel comment they hear, and they won't know if you're joking or not. That's why we had to separate y'all. Every one of you needs to remember you're in the United States Navy and not at home. So you're going to keep your mouths shut and act right. Anyone caught causing trouble is going to have hell to pay. We're not having a race riot on this base just because one smartass can't keep his mouth shut. You got that? Any questions?" our barracks commander asked.

"Yeah. Who's Martin Luther King?" I asked.

"Who're you? Are you Smith? Goddammit, are you really that stupid or are you just trying to be smart? Didn't I just tell you not to joke around with this?"

I dropped my head and said, "I'm not joking, man; I don't know who he is."

"Are you kidding me, sailor; you really don't know who Dr. King is? That's hard to believe, especially here in Memphis. So…how many of you men know who he is?" Three boys raised their hands.

Our advisor shook his head, lit a cigarette and said, "Jeez, three out of forty-four. Where've y'all been, under some rocks? You never heard of the Memphis riots last week and the garbage strikes? The whole world's watching Dr. King right here in Memphis, and you never heard of him? The greatest civil rights leader in the country? In the world? The preacher who tells the Pope and the President what to do? And you've never even heard of him? Well, you will. You'll hear so much about him you'll get sick of it."

"So how long we gonna be locked up in the barracks?" one of the guys asked.

"Don't know," our advisor said, "we'll have to see how long it takes to calm our black brothers down. But do your part and keep the peace. What's it say? Turn the other cheek and love your neighbor."

A bullet had slammed into Dr. King's neck and knocked him to the floor as he walked on the balcony of the Lorraine Motel just after six o'clock. Immediately after the sound of that deadly shot vanished into thin air, a white male initially estimated to be between twenty-six and thirty-two ran from the flophouse across the street from the motel, leaving a .30-06 Remington rifle in its dilapidated doorway. Dr. King died in the emergency room at St. Joseph Hospital at seven.

Tennessee Governor Ellington ordered four thousand Army and Air National Guard members into Memphis from as far away as Nashville. The Governor declared martial law and issued a seven p.m. to five a.m. curfew. He had done the same thing the previous week during a riot near the end of a strike by Memphis' sanitation workers. Dr. King had returned from Atlanta to lead his second protest in support of Memphis' thirteen hundred striking garbage men, ninety-five percent of whom were black.

On April 8th Coretta Scott King and her children, all dressed completely in black, led an estimated forty-two thousand people through the city streets of Memphis in honor of her husband. Mrs. King spoke for about fifteen minutes and urged the tens of thousands of people crammed into and around City Hall to make sure Dr. King's movement never dies. Speaking after Mrs. King, Reverend Ralph David Abernathy sobbed as he told the crowd that his beloved Martin had taken his cross on his shoulder at the Lorraine Motel, and there he was crucified. National Guardsmen stood warily on the city's rooftops, their accompanying helicopters' blades ominously spinning overhead. Rumors circulated through the streets that there would soon be a bloody confrontation between blacks and whites, but on the 16th of April '68, the strike and the curfew mercifully ended without bloodshed when the City finally agreed to the sanitation workers' demands for equal wages and better working conditions. The crushing deaths of two workers in a garbage-compacting accident two months earlier had fed the flames of change into a citywide strike.

About five weeks before Dr. King's death, my fellow shipmate and best friend Peter Gilbert and I had met three college girls in downtown Memphis who were having a beer and looking for something more fun than English Lit. We usually went to town to eat a late breakfast at the Arcade or some fine BBQ at the Rendezvous, but we met these girls sitting at a table on the sidewalk outside a cafe. The girls took us to be harmless but found out we were as serious as rattlesnakes flushed from their den. We were young, brash, full-blooded American sailors with short hair and military IDs to prove it and were looking for some friendly female companionship.

On some weekends, Pete and I would head out to see those young ladies. The Sunday morning after Dr. King had been killed, our classmate Sammy drove us into town to visit them at their apartment. On this morning, we merged onto the expressway heading into town and noticed there weren't ten cars on the road. It had been only three days since the assassination, and people stayed at home watching television; they were scared to leave the safety of their castles, even to go to church. In three or four minutes, an Army jeep carrying three Guardsmen gently forced us off the road and onto the shoulder. When we were completely stopped, they got out and surrounded the car, holding government-issued automatic rifles pointed inside at our bellies.

"Where you boys headed?" one asked.

"Into town."

"What for? Been out all night? Big Saturday night and just getting in? Or going to church?"

"Uh, no, not church." I looked down; I hadn't been to church in two years. "We're going to see some girls. Is that okay?"

"Might be. Don't get defensive, bud. Where're you boys from?" another Guardsman asked.

"We're in school at Millington."

"Oh, so you guys are swabbies. Let's see some IDs."

We handed them our military ID cards, and they slowly looked them over. "Where do your girlfriends live?"

"They're in an apartment at the Hillcrest on East Wind Drive. It's pretty nice. Hey, man," Pete said, "what's going on. I thought the curfew was at night."

"It is, but two black boys were killed late last night by some rednecks in a drive-by, so we're checking everyone." They searched inside the car, but there was nothing to find, except two six-packs of warm Blue Ribbon beer.

"Well, it's a damn good thing you boys hadn't opened one of them beers."

"We know. Thank God, it's too early to drink; I've been in enough trouble lately. Anyway, we've got at least a half hour before we pop those tops," Pete laughed.

"Yeah, you're lucky. Okay, fellows, we've kept you from your fun long enough. But listen up; we could take you in right now as suspects if we were a bunch of jerks, but you're clean. How'd you get off base this morning?"

"Don't know; nobody stopped us. Maybe they haven't heard about the boys getting shot yet; we hadn't," Sammy said.

"Could be; nothing about the frigging Navy surprises me. You're almost there. Go on over to the girls, but I advise you to stay put until this Dr. King thing blows over. No telling what could happen next."

As we drove away, it came to me what had just happened. "Damn, I can't believe they stopped us just because we're white. They were checking us out to see if we were some rednecks with guns going out to shoot some black guys from the other night. Okay, yeah, they have their orders, but this crap is getting scary."

Pete said, "Damn, Smitty, why don't you relax; you're wearing me out, always complaining. They got a job to do, that's all; they don't care about us. They're just a bunch of accountants from Nashville who got called up to play soldier for the week. They're loving it; give 'em a break."

The college girls had plenty of beer in the refrigerator. We watched television, drank a few and talked a little. Two of the girls got up after a while and said they were going outside for some fresh air. The third girl, the prettiest of the three, got up and moved over to sit on Pete's lap. She put her arms around his neck and looked at the other girls. She said, "Y'all go ahead; I'll stay here. I'm going to keep this one."

Unbeknownst to us, in the midst of all our revelry, the lone gunman, the assassin James Earl Ray, was driving his Ford Mustang toward the vast open borders of Canada. After shooting Dr. King, he hurriedly ran out of his flea-bag rooming house and rushed down the street to his Mustang. He dropped his rifle, his binoculars, a radio and a six-pack of beer outside the rooming house's front door. Only a fool would leave his murder weapon in such a conspicuous spot, a place where it would soon be found and with his fingerprints on it, no less. Unless he meant to. He could have taken the crucial evidence with him; he wasn't catching a bus or a cab, he had his own car.

Ray considered himself a smart man, but his life events contradicted that assessment. Through no fault of his own, he didn't begin his life on the right foot. He was only seven when his father passed a bad check in Illinois and moved the family to Missouri and changed their names to Raynes. That incident, although it had no major impact on him at the time, set the tone for his formative youth. Ray joined the US Army as a teenager toward the end of WW2 and served in Germany until he was discharged in 1948 for incompetence and his inability to adjust to military life. Back home, he was arrested in '49 and served two years for armed robbery. He served another four years in Leavenworth when convicted of mail fraud in '55. Soon after his release in 1959, Ray used a pistol to steal a hundred and twenty dollars from a Kroger grocery in St. Louis and was sentenced to twenty years in the Missouri State Penitentiary. He escaped in early '67 by hiding inside a prison bread truck as it left the yard.

On the run from St. Louis to Chicago, then Toronto to Montreal, Ray decided to head south and stopped in Birmingham long enough to buy a 1966 Ford Mustang and get a driver's license. He drove down to Acapulco and eventually settled in Puerto Vallarta in late '67. After an unsuccessful one-month attempt at becoming a pornographic film producer, Ray moved to Los Angeles but soon rented a room in Atlanta. On March 30, 1968, he drove to Birmingham and bought a Remington Model 760 Gamemaster .30-06 rifle and a Redfield 2x-7x scope and some cartridges at the Aeromarine Supply Company.

Returning to his rooming house in Atlanta, Ray read in the newspaper about the upcoming April 1st return of Dr. King to Memphis. On April 2nd, he packed his bag and drove the four hundred miles to Memphis where he took a room at the flophouse directly across the street from the Lorraine Motel. It was easy for him to find the place where Dr. King would stay; it was common knowledge. Reverend Ralph David Abernathy testified to the US House Select Committee on Assassinations that he and his beloved Martin had stayed in that motel's Room 306 so many times that everyone knew it as the "King-Abernathy Suite." Civil rights supporters were enraged years later when Abernathy's biography disclosed the men's activities with groupies in the suite.

On the Thursday evening of April 4, 1968, Dr. King walked out unto the balcony at the corner of the upper level of the motel. He looked around, enjoying the cool breeze, and stood gazing at the waning sunset. James Earl Ray was across the street in the second story bathroom of the rooming house, his .30-06 locked and loaded. He eased the feeble window open and looked outside; everything was peaceful and quiet. The motel's neon sign was intermittently blinking "NO VACANCY" in bright red letters in the twilight, but the parking lot was almost empty; most folks had gone to town in preparation for Dr. King's speech at the Mason Temple.

The Temple was the world headquarters of the Church of God in Christ, the world's largest African-American Pentecostal group. At the

Temple the night before, Dr. King mesmerized the packed audience with his last speech, his ever-famous "I've Been to the Mountaintop." The message presented by Dr. King was a call for unity among the black brethren, a call for economic actions including boycotts against white-owned businesses, and a call for non-violent protests until the City of Memphis changed its course and listened to the concerns of its black citizens, especially its sanitation workers. At the end of the rousing address, Dr. King reflected on the possibility of his untimely death, but he said it was okay; he wanted to live, like anybody else, but he was satisfied if death came, he had been to the mountaintop. He felt he may end up like Moses who was given the honor of leading his people to the Promised Land but was denied the privilege of crossing over into it. If that happened to him, then his best friend Ralph David Abernathy would take the role of Joshua and lead the people to the other side of the River Jordan.

Death threats had been nothing new to Dr. King. He had even been stabbed in 1958, an act upon which he often reflected in his writings. He made clear that any death, including his own, could not stop the civil rights movement. He told Coretta one night after the assassination of President John Kennedy in '63, "This is what's going to happen to me also. I keep telling you; this is a sick society."

Dr. King was standing out on the balcony near his room when the .30-06 bullet tore into him, fired by a gunman who had no job, who had fled from authorities throughout the US, Canada, and Mexico, who apparently and mysteriously had ample monetary funds, and who, it was said, acted alone. The bullet entered Dr. King's right cheek and broke his jaw. It careened down his spinal cord, breaking several vertebrae and cutting his jugular vein and major arteries, until stopped by his shoulder blade. He fell over backward, unto the balcony floor, his tie ripped off by the bullet's power.

Because he still had a faint pulse, Dr. King was rushed to St. Joseph's

Hospital. Doctors opened his chest and performed CPR, but he never regained consciousness and died at 7:05 p.m. Dr. King's biographer, Taylor Branch, said King's autopsy revealed he had the heart of a sixty-year-old man although he was only thirty-nine. He had given his all for his cause; his thirteen years in the civil rights movement had taken its toll.

After he killed Dr. King, Ray fled back to Atlanta to the boarding house where he was staying. A city map he had purchased had Dr. King's residence and church circled in red. Grabbing the few inexpensive pieces of clothing a typical ne'er-do-well bachelor possessed, Ray headed north to Canada. Entering Toronto about the same time Coretta King and her family began their silent memorial march through Memphis, Ray searched out a quiet secluded part of town and rented a room. He hid there for over a month and obtained a Canadian passport under the name of Ramon George Sneyd.

Ray flew to London in late May, drifted down to Portugal, and on his return to London was arrested at Heathrow Airport as he was boarding a plane to Brussels. His Canadian passport with its false name of Sneyd had been placed on a Royal Canadian Mounted Police watch list. The lone gunman also had an unexplainable second passport under a different false name on him at the boarding gate. The wonder of it all was where he found the money to travel all over the US, Mexico, Canada, and Europe. James Earl Ray, a man who considered himself very smart but who had foolishly or deliberately left his rifle, his radio with his old prisoner's number scratched on its side, and his fingerprints at a murder scene, was arrested at Heathrow and was extradited to the US immediately after questioning.

Ray was charged with Dr. King's assassination, and he plead guilty, he said, because his lawyer assured him that was the only way to avoid the death penalty. Bypassing a jury trial, he was sentenced to ninety-nine years in a Federal penitentiary. Ray recanted his confession after only three days and made many attempts to withdraw his guilty plea, but to no avail; he died in prison in 1998. Martin Luther King's family always believed James Earl Ray was just a scapegoat used to cover up a conspiracy between the US government, the Memphis police, and the Mafia, and in '99 they filed a wrongful death lawsuit to explain their positions publicly. The Memphis jury agreed, but in the following year the US Department of Justice reversed the verdict due to a lack of hard evidence.

I told Pete years later, "Man, Ray was just set up at the start to take the fall. He was supposed to leave the evidence and run. Those people behind King's death told him that if he could avoid capture, millions of dollars were waiting for him. Ray was so stupid he thought he could pull off one of the biggest murders in history and get away with it. The cash was sitting in several Brussels safe deposit boxes, they said, but he had to make it there to get it." He almost did; he only missed it by that one final airport boarding gate.

Dr. King's murder was horrible, and our young black sailors had blood in their eyes, and rightly so. The Navy apparently knew what it was doing, even though we didn't. It was a smart move to separate the men by race and defuse any spark that could have led to a bloody battle on the base. All military installations had been trained and prepared for such a situation, and Millington's possible tragedy was handled with the greatest of care and finesse.

Our white sailors were segregated from our black sailors to prevent some smart-ass from saying something racist, giving the black sailors another reason to riot. None of us were allowed to leave our barracks, except to eat under supervision in the chow hall, for two full days. Anyone caught outside their barracks or the chow hall was subject to being put into the brig immediately, but an ass-chewing would have been the real consequence.

It was hard for most of us white boys to understand the problem;

we didn't even know who Martin Luther King was. He had been in Memphis since February; we had been there since January. He had been leading the sanitation workers' strike against the city of Memphis, but we hadn't heard one word about it. Of course we hadn't; that was due to our circumstances. We didn't live in Memphis; we lived on a Navy base. We never saw black sanitation workers; we had our own garbage men, and they were usually sailors, white sailors. We had hardly listened to a radio news report or read a newspaper story on any subject since we had gone to boot camp nine months earlier. We knew nothing about anything in the world. But in our defense, Pete and I were just two skinny white boys in our twenties in avionics school. Radios and newspapers were not readily available at any place near us unless we were off-base, and why in the world would we want to waste our precious time off-base reading or hearing about the local, state, or national news? We didn't realize until much later that Dr. King was a Savior to the black people of America and had been constantly in the news. As his fame grew and his message of non-violence gained credence, that feeling of reverence eventually included most people throughout the world.

Pete and I were typical American kids. We didn't know one thing about anything. All we did was listen to music all night and day and think about scoring with some girls. We didn't know geography, economics, finance, history, or politics. We ran in little groups of like-minded half-wits like a tiny school of fish. Our opinions were the opinions we picked up from our acquaintances. We decided whether we liked someone by their looks and what other equally stupid kids told us about them. We were vain and careless and didn't give a crap about anything except drinking beer and having fun. When we were off the base, radios and newspapers were not on our minds; our obsession was chasing girls in miniskirts, and our lack of success proved it was just an obsession and not an achievement.

2. BOOT CAMP TO MILLINGTON

When Pete dropped out of college, he was unemployed and penniless, just like ninety-five percent of the other flunk-outs. In the 1960s, though, that wasn't that big a deal; eighty-five percent of the kids who graduated were also unemployed and penniless. That's why they went to college. The big difference was only three or four of those other students had run through their father's entire inheritance like he had. Pete and his friends spent every dime on the good things in life; wine, women, marijuana, and food. And the rest of the money was wasted on concerts, cigars, and gambling.

While Pete threw away his four years at Georgia, I wasted my two years at Georgia Tech. I knew Pete in high school at Richmond Academy the one year he was there, but just in passing. He was a senior and a cadet lieutenant in its ROTC program, and I was a sophomore and a lowly private in his company. We spoke once or twice, but no senior wants to associate with a sophomore, and he was a lot cooler anyway. I'm sure that was the real reason, but I made lieutenant, too, by my senior year. I was good at remembering faces throughout my life, but not always a name, which greatly frustrated me. I recognized Pete's face in boot camp the second I saw it.

We happened to be at Great Lakes Naval Station at the same time in the fall of '67. As our separate companies waited in line to enter a mess hall, we stared at each other from twenty feet. We weren't allowed to leave our formation or our place in line, but his line moved up a little, and he stopped right next to me. We laughed, shook hands, and reintroduced ourselves, happy to see a friendly face so far from home. Our companies never had regular times to eat so it was just luck that I saw him. I asked him where he was going after graduation, and he said he wanted to be in aviation. Me, too, I said, maybe I'd see him later, knowing I never would. I didn't notice my line had moved closer to the door, so my company commander yelled, "Cut out that stupid grab-assing, Smith. Get your butt up here before I put my foot in it."

Our paths crossed again in Memphis at Millington Naval Air Station. We were both assigned to Aviation Fire Control School. It wasn't fire control like putting out a house fire, but fire control as in guiding missiles and bombs to an enemy target, thereby controlling the firing of weapons. Naval aviation meant we would be on aircraft carriers and work on Navy jets. We were in the same school, graduated in the same avionics class, and lived in the same barracks. It was a small world, and our true friendship began right there. It was in January of '68, and Pete and I were in our early twenties. It was good to be with someone from home, someone who shared some common memories from high school. When our formal schooling was completed, we were sent to Virginia Beach where we were both assigned to a squadron for the additional on-the-job computer and radar training we would need to join the fleet.

When I saw Pete in boot camp, he had been out of high school five years. It was fun seeing his familiar face; it connected me to the world, freeing me from feeling I was locked up in some prison compound. I bumped into him once in the chow hall, another time while we were getting shots, and once while we marched to class. In boot camp, sailors marched as a group everywhere. In boot camp, sailors were always with their company, always in one unit. We ate together, and we studied, slept, and drilled together. We showered and pooped together.

On a liberty launch heading back to the ship late one night in Hong Kong, a slightly-built sailor arguing with Pete complained about the unbearable conditions he had endured during boot camp. With his typically opinionated and unsympathetic attitude, Pete told the boy, "When I went through boot camp two years ago, it was a snap. If a sailor had the IQ of a Labrador retriever and the body of an ordinary fifteen-year-old kid, it was nothing."

"Give him a break, Pete," I said, turning his attention away from the boy before some other sailor stepped in to defend the boy and tried to whip Pete's butt. I was too tired and too sleepy to get involved with that.

"Sure," he said, putting his booze-infused, red face about an inch in front of mine, "let's give everybody a break. That's how we've ruined everything else. Yeah, some pansy-assed, weak momma-boy like this guy may have had a problem in boot camp, but there were none of those in my company; we ran their asses out. Boot camp ain't nuthin' but a big strainer that captures the psychos and weaklings like this guy, and the deviants and idiots like you, and cleanses them from our ranks; those recruits are not welcome in my Navy."

"You don't have a navy, jackass; you only have you, that's all you care about," I said. "But that strainer crap was pretty good stuff coming from a backwoods Cracker like you. Where'd you read that?"

"I never read a damn thing," he said.

"Man, you think I don't know that?"

Pete smiled and leaned back, affectionately wrapping his thick arm around my skinny neck. After a few quiet moments and just when I thought he had calmed down, the fool suddenly tightened his arm, putting me in a headlock, and laughingly gave me a noogie.

New enlistees going to boot camp were assigned to different companies of around fifty men each. During the four or five months of mental testing and physical training sailors endured, the unfit guys were weeded out, one by one. Night after night sailors went to bed, waking up the next morning to find the recruit beside them gone. The rejects were dismissed silently in the middle of the night or during the day when the company was between drill and classes and barracks clean-up. There was no confusion, no complaining, no one trying to prevent the dismissal. It was all quiet and peaceful. Most observant guys in boot camp could pick out the men soon to be dismissed by watching them. The weaklings and the slow-witted boys were quickly identified and labeled. They had been obviously unfit at the recruiting or draft office but were admitted to fulfill the tremendous quota for fresh recruits required by Vietnam. The psychos came in all shapes and sizes and took a little longer to discover because they mimicked everybody else for a while. But once they were seriously troubled, they went off the deep end, demonstrating intense mental deficiencies and occasionally assaulting someone. Those boys were untouchable; no one messed with them. We waited for the day they were culled from the pack.

But we weren't heartless. We shared a great camaraderie amongst ourselves. We befriended each other and helped other recruits when we could. One boot in our company stood out from all the rest. We were mostly eighteen- to twenty-year-old guys, but this one lonesome fellow looked like he was at least thirty-five. He was of medium height, had a slim build and dark, pointed nipples like a late-developing teenaged girl. His pleasant quietness and friendly smile made everyone like him, but when he hit the obstacle course, everyone knew he was a goner. Our company commander quietly watched and didn't interfere when two or three of us went through the course's physical requirements and then doubled back to help our friend Donny get through his. We verbally encouraged him and urged him on; we often grabbed an arm and pulled him forward, but it was no use. The poor man-boy was a physical failure. He was gone from our company in two weeks. He always reminded me of an old stray weather-beaten dog who comes to your back door and you feed him, and he stays. Like those old dogs, Donny's kind eyes always seemed so grateful for a little love and a small handout. But how'd he get into boot camp anyway? What nut drafted him? No one drafted him; he told me he had volunteered for the Navy because he needed a job and something to eat. Well, this was Vietnam time, and the Navy needed men. It wasn't looking for a few good Marines, it was just looking for acceptable bodies to fill its ranks: warm bodies with maybe a little more brains than brawn. And to its credit it gave Donny a chance.

The sexual deviants were a surprise package. They seemed perfectly normal; they talked and acted just like everybody else, but one morning would find them gone. The scuttlebutt would float around that the guy had been caught during the night soliciting sex in the head, the compartment that contained our toilets and showers, our bathroom. But there was always the question of whether the boy was really gay or had put on an act in order to be discharged, thereby removing his name from the list of men going to Vietnam. Some people thought it was a good tradeoff; a dishonorable discharge versus the chance of getting shot to death halfway around the world. And for what, they'd ask. Many mornings just before daylight, the barrack's petty officer, banging a garbage can and yelling reveille, cut on the lights, and a recruit would hop down from his upper berth to find no one below him. It wasn't like finding an empty, unmade sack; it was finding no sack at all. The missing recruit's bedding and gear had been removed silently during the night. The boy who was approached for sex had reported it to the night watch who told the duty officer who called the guards, and the initiator fulfilled his wish, he was gone. Just like that, quiet and efficient; a section 8 discharge as a Class II homosexual took Vietnam right out of the picture. The beauty of it was the section 8 ruling applied to anyone who was 'mentally unfit.' It was a win-win for the architect of the tactic; he was either gay (section 8) or he was mentally unfit (section 8) for falsely saying he was. What a topic for discussion: was faking a section 8 better than 'tucking tail' and running off to Canada?

Boot camp at Naval Station Great Lakes, a little north of Chicago, ended in December of '67 for Company 359, and we got our first orders. Some of the men were transferred to advanced training schools and some went directly to a ship, becoming members of the traditional black-shoe Navy. Pete and I had selected naval aviation, called the brown-shoe Navy, and were sent to NAS Millington a few miles from Memphis.

Avionics school for Pete and me ran from January through June of '68. It was quite an exciting time to be in Memphis. New students were posted into their barracks as they arrived, and luck would have it that we were both posted to Barracks 38. That was to be our new home for six months.

Sailors and Marines graduated from boot camp at random times, but military classes started at regular times on set schedules. When Pete and I arrived in Millington, we had to wait eight days until the next avionics class began. To keep us busy and out of trouble, the Navy provided us with several meaningful jobs. We were assigned to four different barracks that housed transient sailors on temporary duty, and we had to scrub the barracks' decks and clean the lockers, the beds and the heads each day. There was no use complaining; we had nothing better to do, and no one else was going to do it. We had no choice; months before, at Fort Jackson near Columbia, South Carolina, the captain who swore in recruits from every branch of military service said, "Listen up, men; you are now members of the United States Military, and your ass is mine." Who can argue with that?

Pete and I were caught napping on the job one day and were chewed out royally. It wasn't our fault; we had eaten a big lunch, we were tired from all the scrubbing we'd been doing for three or four days, the sunlight was pouring into the room through the large closed windows, and we only wanted to sit down and rest our legs and eyes for a moment. It wasn't our fault, as Flip Wilson said, the devil made us do it. I'm sure the petty officer who found us dozing had a hard time not laughing his butt off as he roared at us. He probably had a great time telling his buddies how he scared the crap out of two new boots when he lit into us. We were scared; we were in the military during the Vietnam War, and we thought those men had the power of life and death over us, which we found out later they did. We were certainly going to do every single thing they said and right that second, but looking back on it now, there really wasn't much more they could have done to us; we were already going to Vietnam.

Millington was cold as hell in the winter, and we froze our rear ends off waiting to move from one place to another. The problem was we couldn't go anywhere by ourselves; we always marched as a group. If twenty of us wanted some chow, we'd have to stand outside on the sidewalk in formation, four abreast in straight lines, and wait until another ten or twenty guys showed up. Then, after a few minutes at most, all forty of us would neatly and briskly march to the mess hall, led by a petty officer. That was okay in the spring, the summer, and the fall, but it was a stupid rule which froze us to death in the winter. Apparently, the base commander and other officers didn't want two or three thousand young men strolling leisurely up and down the sidewalks between classes, the chow hall and the barracks, grab-assing and sky-larking.

Outside at six-fifteen, we would quietly stand in the twenty-eight degree morning frost waiting for enough guys to come out of the chow hall or the barracks so the group could march to school. We stood there patiently and quietly because no one wants to carry on a bunch of chit-chat at six-fifteen, standing on a sidewalk in the freezing cold, surrounded by a bunch of strangers with your collar pulled high up around your neck and your hands stuck deep inside your peacoat pockets. Inside the classrooms, the warmth from the old furnaces in the middle of the rooms kept everything cozy. The instructors, who were active duty sailors or Marines with years of service, had gotten there before six and had everything ready to start at six-thirty. Pete and I took the same courses, but we weren't always in the same classes. Class hours were filled with reading books and completing quizzes in manuals, and troubleshooting transistors, resistors, capacitors, electron tubes, and on and on. We analyzed, studied, troubleshot, broke down and repaired computers and radars. It was all new and intricate. Had I considered this a part of my instruction for a new and life-long career, I would have found it fascinating. But, just as I had always done, I only did what was needed to get along. All I really wanted to do was go somewhere and do something fun. I just wanted to lie on a beach somewhere, drinking beer, clueless and careless, without a serious thought in my head.

We worked our tails off and welcomed our down time. Not having a car on base made dating hard, so Pete borrowed our friend Sammy's car one evening to take one of our three college girlfriends to a movie she wanted to see. She was his favorite, and she had asked him out. Pete was surprised, actually he was blindsided, when the girl's father showed up at the theater. He was going to be in town on business that evening and had asked her to the movies a few days earlier to see the newly-released and much acclaimed *2001: A Space Odyssey*. She had invited Pete without mentioning anything about her father. The man was a true gentleman, he'd make a great father-in-law, and she was quite a catch, she'd make a fine wife, and Pete got the distinct impression that what was happening was very similar to a shotgun wedding. At the end of the evening, the girl's father shook hands with Pete and told him to take good care of his little girl, making the hair on Pete's neck stand up. Pete and the girl went back to her apartment, and he drank three quick beers while he tried to decipher the evening's events. The beer didn't clear his head, he didn't think it would, and he was tired and confused and a little disappointed that he now had to make a decision about his future with the girl. She had moved her rook and said, 'Check,' and he was going to have to take it with his queen, leaving himself open for another attack. He was too weary to make that move that night.

To keep from falling asleep on the midnight drive back to Millington, Pete occasionally popped his head outside the opened driver's window so the cold night air would keep him awake. Eyestrain caused by an accumulated lack of sleep and his three quick beers made Pete's eyes suffer an intense case of double vision, so he drove with one eye closed on the long, dark haul down the empty four-lane. He assumed his blurred vision was just temporary and didn't worry about it. He had so many other things on his mind he couldn't focus on anything. A couple of miles before the entrance to the base, Pete saw a dog, or a deer, rush from the bushes along the side of the highway and head straight for him. Pete really couldn't tell what it was through his one open eye. He slowly, sluggishly attempted to swerve to the left to avoid the animal, but the animal kept running and was directly in Pete's way. A muffled, crunching noise told him he had hit it. In his daze, he thought he must have hit the poor creature in its chest because it let out a long sighing noise like something had knocked the wind out of it. Pete never slowed down.

The base was abuzz the next day. Word was that someone had gotten killed just three miles from the entrance. Rumors said a black man was found dead in the middle of the highway heading into Millington. The morning news said he was in the middle of the road, presumably trying to hitch a ride when someone had hit him. The police came to that conclusion because his blood was splattered over the fast lane and not in the grass. His car was found parked on the side of the road two miles farther from the base than where his broken body lay frozen in death. He ran out of gas, police speculated, and was trying to get a lift to the barracks. His Navy-blue peacoat and stocking cap naturally blended with the color of his body in the blackness of the night on the unlit road. The police knew it would be impossible to determine if the sailor's death was a result of his skin color or just another accident, but City Hall told them to open an investigation anyway just to keep their political opponents quiet.

Sammy asked Pete what had happened to the front of his car when he noticed a blemish, a small smudge, beside the driver's side headlight the next afternoon. Pete told him he had hit a dog in the night but, thank God, he added, Sammy's old '58 Olds was made of steel. Nothing hurt those things.

Sammy asked, "How in the hell did it hit way up by my headlight? How big was it?"

"I didn't really see it that good. Maybe it was jumping out of the way. Maybe it was a deer and not a dog. I was too sleepy to tell," Pete mumbled.

"Too sleepy to tell? What a bunch of crap. Don't ask for the car again."

Sammy washed and buffed out the offending spot. Pete was right; nothing hurt those big, heavy '58 Oldsmobiles. For a ten-year-old car, it was as good as new.

The young black man had been killed off base, but he was a sailor and belonged to the US Navy. In such cases, jurisdiction was often determined by the authorities of the county where the crime occurred and by the US military justice system. Neither seemed to be particularly engrossed with the boy's death, as several media reports publicly stated. A half-serious investigation, according to some people, was held but nothing was ever found out about the sailor in the road. A sheriff's deputy, speaking off the record, said that was probably because there were no witnesses, no video tapes, no unusual messages, and no evidence of any type. The boy's peacoat had cushioned the force of the impact, but not enough to keep its power from crushing his chest. Unfortunately, the coat prevented the car's paint job from breaking and attaching to the coat or scattering on the ground. It was a hopeless case and was added to the hundreds of other cases in the Cold Case file. Neither Sammy nor Pete were ever questioned, and neither had enough sense to connect Pete to the hit-and-run, or maybe they didn't want to. Ironically, Pete was shown the story in the local newspaper a couple of days later and sincerely said, "If that boy had been white, the Navy would have solved that case. They would have had the FBI in here so fast it would have made your head spin. But a black boy dead in Memphis; nobody around here gives a damn."

A few weeks passed, and disaster struck again. It saddened me, but Pete was surprisingly unfazed by the assassination of Robert F. Kennedy. We were working at completing our avionics courses, so we hadn't been following the election news. A Palestinian Arab with Jordanian citizenship hated Senator Kennedy because he publicly supported Israel. A little past midnight at the end of a long day in June of '68 at a presidential rally in Los Angeles, the Palestinian shot Kennedy three times, once in the head and twice under his right armpit. Kennedy died twenty-six hours later.

"The hatred and the violence will never end," Pete said. "All that great so-called diplomacy is just a waste of time. You have the same thing happening every day in Ireland, in Africa and in the Middle East. There will never be peace on Earth as long as humans live here. Think about the Germans and the Jews." I'd heard it a hundred times and nodded my head in total agreement.

3. MARINE SERGEANT ACKER

Having been to college for a few years and being a little older than the normal recruit, Pete and I enjoyed a little speed and comprehension advantage over most of the other boys in class. So naturally, our avionics instructor Marine Staff Sergeant Acker noticed how quickly and accurately Pete and I completed our lessons and, one day after class, asked us, "What are you two clowns doing in this place?"

Pete said, "We're not doing anything special, Sarge. We're just screwing around, spinning our wheels, and we're gonna enjoy whatever's dished out."

"Were you jokers drafted?"

"No, sir," I said.

"No, sir?" Sarge boomed.

"Do you see any bars on my collar, sailor?" he railed. "Do you see these stripes on my arm? I'm a Marine, a fighting man. Don't call me sir; I work for a living. I'll whip your ass in a minute. Call me Sarge or mister or man. You understand?"

"Sorry, Sarge. I get it," I answered. I thought he was going to hit me. I actually had been slapped in the face by an instructor in boot camp because I left my weapon, my piece, my rifle, but never call it a gun, in the gym while practicing for an upcoming event. It was a mistake I only made once. Of course, Sarge's blowup was just a big show to get his point across: never call an enlisted man sir.

Then, babbling like a fool, I said, "No, Sarge, we enlisted. Pete's draft number was coming up next, so he was about to be called. He wanted the Navy, not the Army. You know how it is; clean sheets and three hot meals a day. All that crap. So he joined the Navy before the Army could get him. My number was nowhere close, but I was just wasting my time, bored to death, I didn't even have two quarters to rub together, so I signed up with him for the fun of it."

"And I didn't want some frigging gook hiding in the jungle shooting at my ass," Pete explained, always having to have the last word.

Sergeant Acker said, "You two lucky pricks; you just may have made the right choice. God knows I've had my ass in a sling many times in this man's war. Look at my arm and leg."

Sarge unbuttoned his sleeve and showed us two jagged bullet wound scars, their deep purple surrounded by spots of white flesh that had covered the places where his meat had been. He pulled his pant leg up, and we gagged at what was left of his calf. He said, "I'm one lucky Marine; I saw some serious shit over there. Man, the very first day I arrived in Nam I was sent to a command post about two miles from our base. I got there just as it was getting dark and reached the door of our bunker just as some sappers cut through the wire. Weapons were popping all around us, and satchels were exploding all over the base every five seconds.

"The major looked at me and said, 'Private, don't just stand there. Is this your first day in Nam or something?'

"I said, 'Yes, sir, it is.'

"He pointed to the bunker's entrance and said 'that's what I thought. Stand right there and kill any and every swinging dick who tries to get near here.' I pulled out my .45 and went to my post. During the next twenty minutes I killed six men, six gooks as you call them.

I'm not proud of that, but I was doing my job, and I'm not ashamed either. They came running at us, all scrunched over, many carrying satchels filled with explosives, trying to throw them in our bunker. One after another. They were littered all around us; the other guys got their share. I thought we were going to kill the wounded, but the word said to keep them alive until someone comes to pick them up for questioning.

"When it was all over, I walked up to the major and asked, 'Is every day like this in Nam, sir?' He grinned and said, 'No, not really.'"

"Is that where you got hit? On your first day?" Pete asked.

"No. I didn't get a scratch that day, but during another attack at a different position about nine months later, some VC crawled through our wire and blew up a small ammo dump. Those fanatics only attack at night so they can't be seen; otherwise, we'd blow them to hell in a minute during the daylight. I slept in my pants and boots just for that occasion and always had my loaded .45 on my chest waiting for the VC.

"I ran to the blast; it was blazing hot, and shells were cooking off every couple of seconds. I saw a Marine silhouetted in the flames, trying to drag a wounded buddy from the fire. I waltzed in sideways, my arm covering the side of my face, when a 105 shell exploded. The blast and the shrapnel threw me outside the inferno but tore my ass up. The other guys didn't make it. So I got some medals for trying to help and a transfer back to the World to recuperate. I requested a school and decided on avionics. I did fine, and now I'm teaching dumbasses like you two idiots."

Sergeant Acker talked to us that way, the way roughneck military men talk to impress underlings and civilians, because he liked us. It was a sign of friendship, a bond between servicemen, welcoming us to the pack.

"Sarge, man, that's awesome. I heard Marines don't like talking about the war and killing people, but we wanna hear your story. You mind telling us?" I asked.

"Well, if you want to. Let's eat chow first and then come back here, say, seven o'clock. If anybody asks, we'll say I had to give y'all some extra instruction. Everybody who knows you two would certainly believe that."

Sarge came back with a brown paper bag with a six-pack inside. "Grab a beer and pull up a chair and 'Listen, my children, and you shall hear...' Ever heard that?"

"Sure, Sarge, it's Whitman or Poe or Longfellow or some other old Yankee about Paul Revere's ride."

"Good job, you're a Rhodes Scholar. An FG."

"An FG?" I asked.

"Yeah. From boot camp at Parris Island. Our drill sergeant always called us a Frigging Genius, an FG for short, whenever we acted like we had any brains at all. Just something he made up," Sarge explained. "It'll never catch on."

"Tell it to us, Sarge."

"Alright, boys, here goes. Remember that major who posted me at the door my first day in Nam? Well, that man took me under his wing; he put me in his command and used me as his right-hand. His name was Pat Burton, Captain Pat Burton at the time. He began in Nam working as an SA, that's a Senior Advisor, in the Mekong Delta, fighting hundreds of dug-in enemies, VC or NVA, whatever, three or four days every week, during his whole tour. Some advisors lived to tell their war stories, but too many don't. The captain's experiences with the Rangers let him see real combat, up close and personal; machine guns, automatic weapons fire, close-in artillery, and mortars, napalm, and mines and booby traps were routine stuff and deadly, and he saw it all from the beginning.

"Pat, he wanted me to call him Pat, said his Ranger days were nothing but deep-water rice paddies and elephant grass, thick jungle and small trails with booby traps, and jungle rot and mangrove ants. He got pissed off when he had to argue with senior officers in helicopters high above a battlefield, who didn't know what it was like to be wading in three feet of water, facing an unseen enemy with heavy machine guns and mortars and trying to evacuate his dead and wounded. But, on the other hand, he said he was inspired by the gunship aviators who kept flying into the teeth of the enemy… and the NCOs who volunteered for Ranger units knowing what extreme hardships and dangers they were heading for… and by the Vietnamese Rangers who overran the best dug-in enemy positions he'd ever seen."

"Hey, Sarge, is it okay to smoke?" I shouldn't have interrupted him, so I smiled.

Sarge stared at me without malice and said, "I didn't know you guys smoked. That crap will kill you before you can give those Viet Cong gooks a chance, and dying from lung cancer will be more painful and take longer over here than a bullet to your head or your chest will over there… But it's okay; grab that butt can; the smoking lamp is lit. You sorry dumbasses are gonna smell my classroom up. I live in this place twelve hours a day, but be my guests."

"Sorry, man, just forget it," Pete said.

"No, I said its okay."

"No thanks. You'd never forgive us, hard-ass."

"Suit yourselves, pukes… Well, Pat arrived in Nam for his second tour right out of the Army Command and General Staff College. That's when I met him at that command post. Major Burton, he was promoted, of course, had inserted a four-man team about six miles northwest of our FSB, a Forward Support Battalion. They reported seeing four hundred and fifty NVA soldiers, or sappers, each with a large box of supplies. Pat went to Colonel Yem, the Vietnamese province chief, to see what he knew, and just by looking in his eyes, Pat could tell that an attack was coming. A full alert was sounded for all the bases near us...You guys still with me? Need another beer? Get one out the bag," Sarge said.

"Thanks, man, keep going," Pete said.

"All our men, even from headquarters, were given ammo and sent to positions on the perimeter. Everybody was going to see some action. The sappers didn't just stroll in when they attacked; we kicked the shit out of 'em with mortars, and gunships and Beehive rockets. Each Beehive round held about a thousand frigging pieces of tiny steel arrows. Night fell, and we were hammered by rockets and mortars. I had a near-miss early in the battle; my eyelashes were singed off by a mortar round. You don't know how much you miss something like that until they're gone. They infiltrated our defenses, breaching the wire in several places, and then fanned out to throw their satchel charges.

"A soldier next to me gunned down a couple of sappers and, when one of those NVA drugged-up maniacs threw a grenade at him, he fell to the floor and covered it with his chest trying to save two wounded GIs in his bunker. But, it was a miracle; the grenade was a dud, so he jumped up and kept on killing the enemy. They broke contact at first light, leaving sixty-three men dead. I kid you not, that battle could have been a disaster for us; thank God the major had sounded the alarm. Two of our boys were killed, and twenty-six were wounded. That morning, when the battle was over, Pat and I examined the dead, and he asked me if I noticed anything odd about them. I said 'yes, they are very dead,' but he meant that they were all Chinese mercenaries; they were way too big to be Vietnamese."

"Did you say some of the dead were Chinese? Are they fighting us in Vietnam, too?" I asked.

"Yeah, but those dead Chinese were mercenaries paid to fight, and not soldiers of a government army. But nearly a thousand uniformed

Chinese soldiers have been killed so far in the war, and four or five thousand are missing in action," Sarge said.

"Okay...let's see," Sarge continued. "Two months later, a Ranger team was inserted at Pat's request on the boundary between Military Regions II and III. Sorry, that's too much; you don't care about any of that stuff.... So, slipping through a screen of heavy underbrush, the five men found themselves on a well-traveled, camouflaged trail. They pushed on and found another and then another. They radioed their troop commander who quickly led a rifle platoon there. When they broke through the undergrowth, they saw that this jungle highway, the biggest trail found so far in Nam, was a giant complex of roads, some wide enough for trucks. Some were hard-packed earth for dry-season movement, and others had year-round bamboo-mat surfaces. It was now clear how the VC down south were getting their supplies.

"We worked hard for the next few months cutting off all the trails in the area, which caused a logjam of the supplies coming down from the north. When the NVA realized they couldn't use those trails anymore, they started transporting the supplies back up to huge supply dumps they built. They couldn't afford to lose all that rice and ammo. Four or five days later, an airmobile scout team told us they spotted a giant supply complex."

"Rice and ammo? That's all?" I asked.

"No, more than that, but that's all those people really needed...just rice and ammo...Pat led us in, and we located the complex almost by accident. I spotted a camouflaged hooch and just followed the bamboo paths from hooch to hooch. We saw street signs and bridges made with walkways and ropes, and what looked like a motor pool and a lumber yard. That was what we called the City, because it was so huge. It was the first of several massive storage areas we discovered. We captured nine new Mercedes, all black, plus a lot of important documents. There must have been a lot of high-up officers and officials living there.

We set up automatic ambushes with mines that killed hundreds of their soldiers as they tried to escape. We used Claymore anti-personnel mines crammed with seven hundred steel balls each, strung in a line to explode in concert with each other. Pat said they were the best weapon we had during the war, and the boy who figured out how to string them together should have been awarded the Medal of Honor. That cache is the largest communist ammunition and arms supply dump captured in the war so far."

"Awesome, man," I said.

Sarge stood up and stretched. He said it was time to go; that class was over, but maybe we'd do it another time. He took the last beer and chugged it. He let out a loud burp and said we were screwing up his life. He would see us the next morning, bright and early, don't be late. I told Pete that Sarge had taken a shine to us. We straightened up the place and took the beer cans and the bag with us. It was the least we could do.

On several weekends, Sarge drove us to Blue Lake, Arkansas to swim in a large, pristine lake favored by military families and to drink beer all day under some shady shoreline trees. Pete and I saw plenty of Sarge in and out of class and enjoyed his friendship until we eventually had to move on. One special weekend in May, Sarge rented a cabin for the three of us five miles above Blue Lake. We were going to stay overnight. Pete and I weren't worried; there were two of us and only one of him, just in case anything unusual arose. We left early in the morning before sunup, which was the same thing my daddy always did when he took me fishing.

We drove through the black section of Memphis on the outskirts of town, pulling his small Jon boat behind his big, virile pickup truck which was tough and sporty; almost the exact opposite of him. Sarge cut his lights off as we coasted through the neighborhood, and I asked him why. He said it was to keep from bothering all the sleeping folks. We were quiet as a mouse on the smooth, packed dirt of the road. He pulled up at a dark screened-in porch behind a white picket fence and gently knocked on the door. The old black lady who lived there got out of bed and showed us all her different types of bait. It was Saturday, and she was expecting him, so it didn't bother her at all. That's why she was there, she said, she needed the money. We were gone from her place while it was still dark. Off base, Sarge acted like a good ole country boy, quite different from his northern Jewish upbringing, and all the black folks liked him because he was pleasant and fair with them. He treated them with kindness and respect; he treated them like family. He was beginning to remind me of my daddy. There was a certain peacefulness as we gently rolled along the quiet, dark unpaved streets heading west out of town.

After thirty minutes or so of complete quietness, I saw Sarge rub his forehead and sigh. "Everything okay?" I asked.

"Yeah, I was just thinking about the war. Did you guys hear about the Tet Offensive over in Nam? Started this January and is still going on. I wish I was over there helping my boys. It's pretty bad; I know I lost some buddies. And two Marine regiments are fighting on a hilltop called Khe Sanh. It started with Tet and their base was surrounded until just a couple of weeks ago. No telling who made it out of there; I'm waiting to hear."

Unbeknownst to Sarge, something possibly more horrible than Tet and Khe Sanh was happening over there, but no one knew of it except the soldiers who attacked and killed the civilians of a hamlet called My Lai in South Vietnam. It would take 20 long, hard months before the entire world would know.

Some other long-time friends of Sarge were in two of the cabins next door when we arrived. They already had their full-sized boats in the water, but the beautiful thing was that the river where we were was only fifty yards wide, so our little boat was just as good as theirs. The land was swampy and umbrellaed with trees where it flowed into the pristine lake, so the river water was that clear brown, acidic color of a glass of diluted Coca-Cola.

We unloaded Sarge's pickup and put his little Jon boat in the water. Sarge's friends loaded us with water and snacks. The three boats headed north on the river, and each of us stopped to tie onto a tree branch along the riverbank after a mile or so. We fixed up our poles and dropped our lines into the water. We were ready, and we talked about everything as we waited between bites for another fish. Sarge looked around us at the trees and bushes and the few fields out beyond the river and unto the horizon.

We fished til noon and caught a mess. We untied from our tree branch, and our Jon boat's tiny little engine tooted us back to camp. Sarge's friends fixed us some lunch, and we drank a few beers. We pulled the six-foot square by three-foot deep skeleton cage holding the fish from the river, shook them out, cleaned them and put them in a cabin refrigerator.

Near the banks we could see the white sand on the river's bottom from the side of the boat. We could see the movements of all the water animals and fish silhouetted against the sandy bottom. We fished along the sides of the little river near the bank with reinforced cane poles and worms Sarge brought for Pete and me. Sarge had an expensive, professional rig, as did the other guys, and used lures shaped like shiners, worms and other bugs. Pete and I caught almost as many fish as they did. There's nothing like pulling a big ole crappie or catfish or trout out of the water and making that cane pole almost double up into a perfect upside-down U. The river was beautiful, and the day was still warm. The only worry we had was to make sure a copperhead was not sunning on a branch we had selected to tie up to as we moved along the river to different fishing holes. Sarge kept a pistol in the boat and had enough sense not to shoot anything inside the boat. We saw many snakes, but they left us alone, and we left them alone.

Around three o'clock, in the anticipated coolness of the afternoon, we launched our boats again, heading south toward Blue Lake this time. This wasn't Irene's first trip with the boys; she had been on many of them. Jack was a Bassmaster and only did two things; he worked and he fished. Irene accompanied him everywhere, so everybody knew the routine when Irene had to answer nature's call, and it waited about two hours this time. It was no surprise; she had to go pee just about every time she went out in a boat. It may have been because she and Jack usually drank four or five beers each to calm their nerves, they said, as if they had anything on earth to worry about. There was no ladies' room, and that afternoon she chose to use the woods and not a big-mouthed Mason jar. To keep from scaring the fish, Jack gently paddled his boat closer to the water's edge. Irene grabbed the first branch in range and pulled the boat right up to the bank. The sides of Jack's new boat were higher than those Irene was used to, and the side of the boat interfered with her footing as she climbed over. It stopped her momentum halfway out of the boat. She had one foot in the boat and the other foot on the bank, and she couldn't move.

"Help me, Jack! I'm going to fall, dammit," she cried.

Jack was in the back of the boat near the motor, and the other two guys were down stream, so no one was there to help Irene. We were only twenty yards down the river but just sat there, watching the adventure unfold. We looked at Sarge, but he just quietly shook his head and smiled. He had seen Irene's dramatics before and wanted us to witness them for ourselves.

As Irene tried to decide what to do, whether to start over or to try to make it to land, the small dried and withered branch snapped off in her hand. With nothing holding it to the bank, the boat now began to slide away from the shore with the current. Irene had waited too long to act, and she was doing a split. Her legs kept getting farther and farther apart. She leaned towards the shore, but it was gone. Had she leaned back into the boat, she may have been fine. Instead, she flipped over backwards into the water with a loud howl. Everybody, including Bill and Howard, had been watching her dilemma and listening to her squeals and roared with laughter when she disappeared from view. When she popped up and regained her footing, she found she was standing in only three feet of water. She laughed and looked around and yelled to us all, "I might as well pee while I'm in here."

Everybody drank and drank, and ate and ate, when we came in. There is nothing better than frying fish caught that day and eating them with grits and tomatoes and pickles and white bread. Sitting about twenty feet from the red-hot Coleman stove under the tin-roofed shelter beside a riverbank reminded me of some good times back home with my daddy. After all the fun and excitement of the day, Pete and I finally went back to the room we shared and fell asleep. Apparently, no one else did. The next morning, we were awakened around nine when one of the guys stopped by to say goodbye. I opened the door and he asked where Sarge was. I glanced around the room and said I guess he was in the bathroom and I'll tell him you said goodbye. Sarge was terribly hung over and fell asleep naked on the john, and after I woke him the second time, he finally showered and got out. We were the last to leave. Pete drove Sarge's pickup as we headed back to the base. I was looking forward to getting back to the peace and quiet of Millington.

4. FOGGY WINDOWS

We graduated from our avionics school near the end of June of '68, the first major step in our naval careers after boot camp. Pete and I were both transferred to VA-42 at Naval Air Station Oceana, Virginia Beach, Virginia. We had a one-week leave before reporting to Oceana, and we both went home to Augusta, but not together.

I hung out with my family in Augusta for a few days while renewing a friendship with an old girlfriend who had said she would always be my hometown girl. Pete didn't have an old flame to hook up with; they were all married or knocked up. Between college and the Navy, he had only bothered to go home a dozen times in five years. My girl lined Pete up with one of her buddies, and they hit it off well, so we double-dated a second time on Saturday night.

Dating in the mid '60s wasn't very complicated. During our high school years, we usually went somewhere and drank beer in the car at a drive-in movie, or on a dirt road in the country, or on the steps of a mausoleum and listened to the radio. I turned twenty-one in February of '68, and Pete was twenty-three, so we were plenty old enough to drink, but the girls were only eighteen. The legal drinking age in Georgia and South Carolina was twenty-one at that time, but Pete knew a place over in South Carolina where he had been drinking beer since he was sixteen. He said the girls would be fine there without showing any

IDs. I drove the girls over to Pete's house in my daddy's little sports car, and we borrowed his mother Midge's car with the big back seat.

We walked down the stone stairs to the entrance of the club, and of all the dirty tricks, there was a bouncer at the door. We decided we had already driven across the river, so why not see what the man says; we had nothing to lose. He looked at our military IDs and said, "Hey, you guys in the Navy?"

"Yeah, we're sailors...home on leave. That okay?" I asked.

"Sure, man. We get soldiers in here all the time from the Fort, but never any sailors. My brother is on the *Wood County* right now. I got to go on board last summer in Norfolk; you ever seen it?"

"Nope, but we're stationed about twenty miles from there."

"Cool. Come on in, but soldiers drink here all the time, so don't start any trouble. Okay?"

Without even glancing at the two young girls, he said, "Come on. Let me clean a table for you guys."

I asked what the cover charge was, and he said, "Forget it, man. It's on me. Have fun."

We were in civilian clothes, and a couple of soldiers from Fort Gordon were there in their neatly pressed, handsome uniforms. Pete kept looking at them during the hour we sat there laughing and drinking. They must have been boots, because they had closely cut heads and no medals. Everybody in the service said the Army was handing out medals just to keep the troops happy so they'd re-up for Vietnam. The soldiers were offered a little more pay, and better jobs and safer environs, but if medals were the best way to win their hearts, let it be. So, if a soldier didn't have any medals, he was a new recruit, and that was fine; we were basically new recruits ourselves.

Pete drank too much as usual and after a while started acting like the jackass he was, talking trash to the two soldiers who had been nothing but perfect gentlemen. Of particular interest to Pete was one soldier's spit-shined dress shoes; each toe was polished to a mirror-finish. Pete said to the man, "You guys in ROTC at Richmond Academy?"

The soldier flinched and said, "What?"

"I had a pair of spit-shined shoes like that in high school at Richmond Academy," Pete said. "Actually, even before that, at Gordon Military. I thought you boys might be in ROTC, but why would you wear those uniforms in public?"

The soldier glanced at his buddy and started to rise. I wasn't about to fight those guys, especially with two dates at our side. Raising my open hands, I stood up and said to anyone who would listen, "Hey, stop! We don't want any trouble. I'm sorry. My friend has had too much to drink, and I apologize for him. It's time for us to go anyway."

Then I lied, because I knew what a cruel mean streak Pete had, and I knew he wasn't joking; he was drunk and was trying to start a fight. "Sorry, guys, he was just trying to make a stupid joke. Please excuse us. Get up, girls, we're leaving."

Heading to the car for the trip home, Pete called me four or five bad names, but then asked me to drive; not because he was a little drunk, but because he wanted to make out with the new girl. It had gotten dark outside, but I could still see he was all over her in the back seat. She'd had a few beers and was having a ball, so I didn't say anything except, "Damn, Pete, most people wait until they've parked somewhere before they get into all that stuff."

Twenty minutes later, I pulled his mother's '64 Chevy off the paved road five or six miles from the girls' neighborhood. I eased down the dirt road leading into a newly surveyed subdivision. I carefully drove along the lots with survey markers attached to metal poles where the new houses were soon to be built. It was dark, and there was no moon, making it hard to see the markers fluttering like military flags standing at attention on a parade route. It was a perfect night to be in love, but I didn't know where the hell I was. I turned into a soon-to-be driveway, circled the car around, facing it toward the main road running down the center of the subdivision, and cut it off.

"Come on, sugar," I said to my girl, "let's go for a walk. We'll be back in an hour, so don't go anywhere, you guys. Have fun."

We got out so Pete and his new friend could have some privacy, if that's what they wanted. I knew for a fact that's what they wanted, so I grabbed a couple of the cans of beer we bought in Carolina, and we walked to the trees behind the car. The night was delightful, a little cool; a million stars were visible.

"You okay?" I asked.

"Sure," she said, "but I'm going to miss you when you're gone."

"Me, too, but we still have four more days. Can you drive us to the Columbia airport when we catch our plane?"

"Of course, I'll get the information tomorrow, but I was thinking about our future. When am I going to see you? I'll be in Athens this fall, but I can fly up there anytime," she said.

We'd never talked about a future before, so I didn't know what to say. She was the sweetest, cutest thing. "Sure, I'd love to see you at Oceana, but I'll have a lot of work to do, and we'll start going to sea in eight or nine months. That would give us a long time to go steady, but do you want to have a long-distant romance while you're at Georgia? That wouldn't be much fun, would it?"

"We can date other people and still stay close to each other. I've loved you for a long time, and you weren't always there. Let's don't quit now; this has really been fun for me. What do we have to lose?" she asked.

I bent down and kissed her beautiful lips and eyes. She was the best girl I had ever dated. Anyone would be proud of her, but I didn't want to get too involved with one person so far away. Having a long-distant love affair would be just as hard for me as for her. Long letters, late-night telephone calls, occasional visits; who really wanted that?

"You're right. We have nothing to lose. Let's see what happens," I said.

"Do you love me?" she asked.

I hesitated, but I knew what I had to say, "Of course I do, you know there's nobody else but you. And I always will."

Well, that was close and necessary and diplomatic and a little ambiguous. I hadn't been in love before, so I didn't know what to say, but I knew not to say no. Many guys talk about screwing all their girlfriends and never telling them they love them, but that seemed to me to be so cold, so uncomfortable and maybe even mean. They could at least make an effort to please the girl, even if they had to tell a little white lie. Saying you loved them didn't mean you had to marry them, did it? But sometimes it was a good start. So you break up soon; big deal, what did it hurt to make her feel good?

Time came to get the girls home. As I approached the car, I noticed Pete and his girl had really fogged up the windows. It looked like they had kept them rolled up all the way the entire time they had been in there. I wondered why they hadn't suffocated.

The windows were so heavily fogged I couldn't see a thing inside. To make sure they were decent before I violated their privacy, I knocked on the back window. I knew Pete didn't care if they were decent or not, and neither did I, but I didn't know what the eighteen-year-old girl thought. When Pete rolled the window down, a blast of warm fragrant air oozed out like smoke from a dying campfire on a cold night. It seemed thick enough to see. I was momentarily taken back by the power of it.

"What y'all been doing?" I joked. "It's time to get home. You ready?"

"Sure," Pete said, "we've just been lying here waiting for you."

"Well, I don't know what the hell y'all been lying on, but it smells worse than Vignati's back there."

"Vignati's?" Pete asked.

"Yeah, the fish market on Broad Street. You never walked into that place? Good God, it stinks. You can smell it three doors down, outside on the sidewalk. It smells like somebody with a bad upset stomach has been passing gas, but it does have the best fish and lobster tails in Augusta," I said.

"Ha-ha," the girl scowled as she straightened the front of her dress, "you're a real comedian!"

"You know, it does smell like fish in here," Pete joked.

Pete and the girl were still sprawled on the back seat, and my girl and I hopped in the front. I started the car and let it slowly creep forward toward the main dirt road. Blind as a bat, I kept moving forward, slowly forward. There were no streetlights at any place in the new development. The windshield was so fogged up not a thing could be seen. I flipped the switch to cut the defroster on high. The whooshing noise from the defroster was loud and high-pitched, but I still couldn't see. If I had had any sense, I would have stopped the car and waited until the windshield was clear, but no, I had to keep on going. I hoped there were no people or animals ahead of us. How could there be; nothing was out there but stacks of bricks, sacks of concrete, two by fours and foundation holes.

I tried to wipe the gray cloud off the windshield with the back of my hand, but its heat made it worse. We rolled about twenty feet, and the car suddenly started leaning forward. What the hell, were the front tires flat? Did someone come up to the car while Pete was erotically engaged and cut the tires? Were there some men out there with knives, or guns? I gently pressed the brakes, but the car had already stopped by itself. Had I hit something? Had I hit one of those men, the pranksters, the robbers, the rapists? Why hadn't Pete or I brought a pistol from home?

I thought I'd better back away from whatever it was, so I put the car in reverse and gave it some gas. Nothing happened; the car did not move. I shifted back to forward and tapped the gas. Nothing. I tried reverse one more time. The only response to more gas was the RPM needle jumped up like we were going somewhere, but we weren't. The roar of the engine let me know it was fine; nothing wrong there.

"What the hell's going on?" Pete asked. "What's the matter?"

"Who knows? Get out and see," I told him.

I turned the key which cut the engine off. Pete and his girl got out of the car. I handed Pete a flashlight I found under the front seat. Thank God, Pete's mother had put it there.

Peter Boy yelled, "The car's suspended in air. The back wheels are floating about half a foot off the ground. You're in a hole; a foundation hole for a house. And the front wheels aren't touching anything, either. You're like a see-saw floating on air."

I said, "Oh, God! Are you kidding me?"

"No, I'm telling you the truth. You're balanced like a see-saw."

I asked Pete, "Can you rock it or sit on the back bumper and make the back wheels touch the ground?"

"I'll try," Pete said, "but I don't want to hurt momma's car. You do whatever you need to, though, she won't get mad at you."

"Alright, I'll be careful," I said, "but we can't stay here all night. Whatever you do, do it slowly and try not to make the car slide forward. I only want to go back where we were."

Pete went to the back of the car and sat on the very end of the trunk. The back of the car did move down a hair, but it was not enough. The car was perfectly balanced on the edge of a giant professionally-made hole. Pete called his girl over and made her sit on the trunk with him. That was better, but it still didn't do the trick. He made her gently bounce up and down with him, but it scared the hell out of everybody, and they stopped. Unfortunately, the wheels never touched the ground.

The time came when I thought I'd have to walk the two miles back to the convenience store we had passed, hopefully it was still open, and call the police, or a wrecker, or even worse, Pete's mother or my father. It was too late to knock on someone's door along the way; it would scare them to death. Being in Georgia, in the country, I was more afraid they'd shoot me to death at that late hour, standing unannounced at their door in the middle of the night. But at that moment Pete did something I'll remember 'til the day I die. He walked around and dropped into the hole in the front of the car. He was big, but the hole was much bigger. It was at least twenty feet wide and fifty feet long. It was four feet deep. It was the beginning of a fine crawl space or a basement for a soon-to-be built house.

Pete looked up toward my direction and yelled, "I think I can work this out. Let me get the jack out of the trunk."

He hopped out of the hole, carefully popped the trunk and took out the jack. He walked back around front and jumped down into the hole. He gathered five cinder blocks he found inside the hole and stacked them underneath the front of the car. Two blocks pressed side-by-side made the base, another two turned at ninety degrees formed the second tier, and the fifth topped the pyramid. Pete then set the jack on that top block, aligned it beneath the car where it seemed the strongest, and began pumping its arm up and down, up and down. The front of the car inched upward slowly which, of course, pushed its back downward. After about five minutes of solid pumping, the tires in the back touched dirt. Pete kept pumping until the girl said the tires were solidly on the ground.

The girl said we were ready, and I yelled to Pete, "Get away from that jack. I'm going to back us out."

I yelled to Pete's girl, "Sally, get away from the car."

I cranked the car up and slipped it into reverse. I didn't know whether to gun the engine to get us out in one quick surge or to slowly back out. I chose to take it easy, and the car slowly responded to my gentle touch. It backed out as if nothing had ever been wrong. As the car moved farther from the jack and its pyramid of cinder blocks, the top of the jack supporting the car was pulled toward the edge of the hole. It reached its physical limit, and then with a loud clank it flew away like an arrow sprung from a bow and lodged itself into an earthen stanchion in front of the car. Pete grabbed the jack and put it in the trunk when he got out of the hole. He brushed himself off and stood outside as Sally got in. The windshield had finally become clear, and when we were far enough from the hole, I put the car in Drive. I asked Pete if he wanted to get in and drive. He yelled, "Hell, no!"

I told Pete to lead us out of there; we didn't want a repeat performance. He walked in front of the car and led me out. He got in, and we took the girls home. It was one thirty-five. I hoped the girls didn't get into too much trouble, but sometimes that's the price you have to pay for true love.

I didn't see much of Pete during those last few days of leave because I wanted to spend more time with my girlfriend. I was going to see three more years of Pete, but who knew when, or if, I would ever see her again. She said she would wait for me, but she'd be a fool to do it.

The Navy had arranged free passage for Pete and me from Columbia to Norfolk. That sweet girl drove us into the middle of Fort Jackson in her little Corvair, and an Army bus took us to the airport. Pete and I were on the same flight, and our spirits were sky high. Every seat on the charter flight had a newly minted Marine or sailor in it, and we drank and played poker. My old man would have been proud of my gambling success.

At the Norfolk airport, we unloaded our duffel bags and were led to a Greyhound bus which left an hour later for NAS Oceana. We were excited about Virginia Beach, even the name was cool. The World's Largest Resort City was the third largest city in Virginia with 175,000 residents and thirty-eight miles of waterfront and bays. We found our barracks, unpacked and went to chow. Oceana's twenty-five-year-old fire-traps were similar to the ones we had in Millington, but it didn't matter to me or Pete; we were on an adventure. We were young and single, we were good-looking and smart, and we were in the US Navy without a worry in the world. At that time.

Our first assignment at Oceana in July of '68 was to Replacement Air Group squadron VA-42. Called a RAG outfit, the squadron trained Navy aviators, flight officers, and aircrewmen for their future permanent duty. Pete and I were enlisted aircrewmen training to install, operate and repair the computer and radars in the Grumman A6A Intruder All-Weather Fighter/Bomber. We had already passed the basic courses of computers and radars in school at Millington, but VA-42 would give us the on-the-job training we'd need in the fleet.

We'd been studying computers and radars for months and thought we were something special. We worked on sophisticated jet aircraft that carried the finest electronic gear available in the world. We were told our computers were more advanced than the ones used in the nation's space program and in satellites circling the Earth. A large, screened-in space in the hangar was reserved for us when we were home. We called it our shop, and our equipment and spare parts were stashed all over it. We would meet in the shop and discuss what was needed to get the system back up after a mission debriefer told us what complaints, if any, our aviators had with a plane after a flight. Gathering our tools and maybe a computer part or two, we walked into the hangar and started analyzing and repairing our plane. The most important thing we were taught was to make sure safety pins were inserted into the ejection shells on the aviators' seats. The second rule was to never hit the Transmit switch for our two radars while we were in the hangar. Each A6 had a search and a track radar; the search radar displayed far-away land features and objects on the bombardier/navigator's screen, and the pilot used the track radar to lock-on and destroy enemy aircraft and ground targets. Those radars sent constantly-updated information to the computer to control the plane's speed and direction on missile strikes or pin-point bomb releases. The Navy was afraid we'd cook somebody if we splashed them with radar waves at close range.

We were special indeed, so it came as a big surprise when one day we were ordered to walk up and down the flight line and all around our hangar to pick up tiny pieces of debris that could be sucked into a jet engine.

I asked the petty officer leading us along the flight line, "Chief, what are we doing out here picking up this crap?"

"You're doing your job, sailor."

"But why did the Navy send us to all those schools, just to have us out here doing a job that anybody can do?"

"It's the Navy's way to keep you humble, boy. Pick up the pace; you're falling behind."

"Aye, aye, Chief."

That was the Navy way; you were told to do something, and you did it. You might bitch about it a little, but it had better be only a little. I carried that lesson with me the rest of my life. The petty officers who led us taught us many things, and we appreciated their help.

Life was fun for young white boys in Virginia Beach in '68. Pete and I gave up our initial desire to have an apartment and decided to settle into the barracks. My girlfriend, now a Georgia coed, came to see me on occasion and always invited me to stay with her at a motel on the beach. She loved the clubs that played pop music, especially a local nightclub called Rogue's Gallery, home to Bill Deal and the Rhondels until they began touring nationally. Virginia Beach had other night clubs such as The Mecca, The Golden Garter and Peabody's. The Civic Center featured acts like the Rolling Stones, Jimi Hendrix and the Grateful Dead. We drove to the coliseum in Hampton Roads to see Dionne Warwick at her peak in '69. We stayed in touch with the music scene the entire four years we were in the Navy. We looked and acted differently on the outside, but we hadn't changed one bit on the inside.

In the clubs and on the streets, it was obvious to everyone we were sailors. We had nice, neat haircuts and the other guys had Beatle cuts, or worse; they had that long unkempt sun-bleached surfer mop. But the civilians couldn't tell we were sailors by our uniforms, because we didn't wear uniforms when we were off-duty. It wasn't that we weren't proud of them or of being in the Navy; the problem was we weren't allowed to wear our more casual Navy work clothes off the base. Every time we left the base in uniform, Navy policy said we had to wear our seasonal dress uniforms. That meant we had to shower and get redressed, and once we had showered it was easier to put on civilian clothes than uniforms. Besides, most of us had our dress uniforms professionally dry-cleaned and pressed for inspections, and we couldn't afford to mess them up eating pizza and spaghetti and spilling drinks on them while partying.

The policy was fine, and we understood it. We worked around jet fuel and grease, dirt and dust, in hangars and on the flight line. The Navy didn't want us going off the base interacting with civilians while looking like a bunch of dirty kids who'd just crawled out of a mud puddle. We were to present a good image of the Navy at all times. Pete and I didn't mingle with the public often, so we didn't care. Very few enlisted men on base had civilian coats and ties, so we saved our dress uniforms for very special occasions. But while we were at Oceana, acting under pressure to keep the Vietnam recruits happy, the Navy decided to let us wear our work clothes to and from the base, as long as the clothes were clean and presentable, and we didn't go drinking or shopping in them.

Pete and I went to dance clubs as money permitted and always in civilian clothes. Some females frequenting the clubs liked personable sailors; some didn't, and those stuck up women would not patronize clubs with too many sailors. Some run-of-the-mill local men were naturally jealous of sailors who had something on the ball. Club personnel noticed fewer local men in the clubs when a lot of sailors were there. More than once we got intimidating looks as we paid our cover to get in. The unwelcomed reactions at the local night clubs was nothing personal, it was a ratio problem. The more sailors in the club, the fewer girls to go around. We got it, but it shouldn't have even been a contest; we only had enough money to get into a club and buy a couple of beers, and we lived on a naval base. Anyone working outside the Navy had the advantage; they had more money to spend on the girls and a place to take them after the club closed. But the girls liked Pete and me.

At the time Pete and I were in the service, there were no female sailors in the avionics branch of the Navy. I cannot remember seeing any female servicewomen, other than the ten I may have seen on a base, during the entire four years I served, probably because women were not allowed to serve on ships. Aircraft carriers had no women on board; the powers to be wanted them separated from the dangers of life aboard a combat ship. They were also aware of the potential trouble of having women in close contact with men who hadn't been intimate in months. A few women around thousands of sailors on a confined ship could be problematic. Relationships could prove deadly on military ships, just as in civilian life, when someone makes a partner change, and the other partner is not willing to accept the change. Jealousy breeds thoughts of revenge sometimes in the minds of people who have been replaced and cannot escape the environs where their past lover works. Women were first allowed to serve on non-combat ships in '78 and on combat ships in '94.

From the male perspective, the principle of wooing females applied throughout the military: make them feel special even if they, unfortunately, happened to be prostitutes. In every country where we went ashore, American servicemen never took the presence of any woman for granted, and if they found one they liked, they tried to make her happy. Sitting on a patio overlooking a beach in Greece or drinking in a bar in Japan or the Philippines, it was all the same; four or five sailors would be crowded around a table, laughing and acting like fools, trying to gain the favor of a woman they thought was on vacation, but who was usually a hooker. Thirty-five-year-old women who go on vacations, or holidays, by themselves are usually interested in more than sight-seeing; they are usually on the clock and looking for cash. But that makes no difference to sailors; these women were on a special pedestal, they were women. The scene usually acts out like the plain, colorless female Redbird being courted by five or six beautiful dark red males looking for a little love. She only picks the best when there's more than one, and in opposition to her human counterpart, offers him what she has for free.

    I told Pete I wanted to work the night shift at Oceana so we could go to the beach during the day. That meant we wouldn't be out drinking and partying and spending our few precious dollars every night. He said that sounded good, and we both were put on the three p.m. to midnight shift. It felt a little weird getting off work at midnight, but we quickly adjusted. We ate lunch in the chow hall just before it closed at two o'clock in the afternoon, and then we'd go back for our supper around six-thirty that evening. We'd walk over to one of the fast-food places after work for a snack. If you knew what you were doing, you never had to go hungry in the service. The chow hall was never actually locked, and a petty officer was always there with the night crew getting the place ready for the next morning. There were floors to mop, tables and chairs to clean, pots and pans to scrub, condiments to refill, the next day's meals to organize and ingredients placed at hand for cooking. The sign on the door said closed, but that wasn't necessarily true.

    Men on watch at Oceana had to have something to eat before they headed to their duty station and afterwards. If they wanted, they went to the chow hall and were given a meal. At any time. But you had to know to go there and say you were going on watch or had just come off watch. The sailor on duty would take you to a giant walk-in refrigerator and while you stood there answering his questions, he would fill your tray or napkin with the leftovers you requested. Cold fried chicken and ham or bologna sandwiches piled high with lettuce, tomato, and onions were my favorites. After we got a bite to eat somewhere, we'd head back to the barracks, shower and get some sleep.

Around nine in the morning, we'd be up and about, finished with breakfast and our personal errands, and hit the beach. We usually took the base shuttle into town and squeezed between the motels to get on the beach. We'd lie in the morning sunlight and then the intense noon-day sun. We had the best tans of anyone in Virginia, resulting in my daily application of skin cancer cream fifty years later. Of course, it was too early in the day to pick up girls, but during the week there weren't any to pick up anyway. In fact, Virginia Beach was a great family beach where parents took their children on safe, quiet vacations. Half the visitors came from Virginia and Canada. So, if you wanted to pick up women, you generally had to go to a club at night or to the beach on the weekends, or to Myrtle Beach or Florida.

5. VIRGINIA BEACH

I met someone new that fall. Pete and I had been drinking over at the enlisted men's club one evening and decided to thumb to Washington, DC to see a girl friend of mine. We were going to thumb to DC during the night, and we didn't have sixty dollars between us. You would understand it if you had been a twenty-one-year-old boy who was looking for love in all the right places, and occasionally in all the wrong places. I should say white boy because I'm not sure a twenty-one-year-old black boy would have arrived alive had he been thumbing through Virginia on the interstate highway in the middle of the night in '68.

We were working days, and it was seven-thirty one Friday evening. I had been talking by phone to an old friend over the past three months, but I never said I'd go see her. She wasn't a girlfriend; she was a girl friend, someone with whom I'd never been intimate. I called her from the club, and she said for us to come on up. It was only a four hour drive, but we didn't have a car, so we were going to thumb from Virginia Beach to Washington, DC. I know it sounds incredible, but we were pretty boozed up. We also kept drinking while we waited for each new ride on the road. We had with us two very large bottles of cheap wine, each wrapped in a paper grocery bag. In addition, we both took a clean shirt and a pair of underpants and a toothbrush and paste in one cheap rag-tag gym bag. When our third ride let us out south of

Richmond, we had already been on the road five hours. The next ride picked us up after we sat on the I-95 guardrail for four more hours. He was going to Silver Springs, MD. We thanked him for the lift, talked for fifteen minutes, and passed out. He woke us up around six o'clock and dropped us off at DC's edge. He had gone out of his way to get us inside the Beltway, close to downtown. We told him how indebted we were, and we were dead serious.

We caught a city bus and rode with the city's morning hotel staff and restaurant workers to Connecticut Avenue. Dunbarton College of the Holy Cross was an exclusive and expensive all-girls Catholic school. I knew my friend from high school was rich, so I assumed all her fancy friends were, too. Pete and I were young, handsome and eligible; we didn't mind if they were rich elitists. We walked to the school and waited until breakfast before we inched our way inside. I called my friend from a pay phone in the lobby. We still had our travel bag and one unopened bottle of wine in its Piggly-Wiggly paper bag. Our sophistication oozed from us like the odor of the cheap bitter wine residue seeping from our unbrushed teeth and mouths.

It was Saturday, and almost all the nuns were off duty. My friend had a date that night that she couldn't break, so she said, and set me up with one of her roommates. I didn't get too close to her until I had cleaned up a bit. A friend from Oceana had a brother at Georgetown University with whom we had met and partied like fools at the beach. We had shown him a great time, and he owed us one. We called him and he came and took us to his dorm where we showered and put on our clean shirts and underwear.

We took the bus back to Dunbarton where the girls had stashed our wine. The new girl was very pretty and smart. I enjoyed her company and never forgot her. We toured the city, ate the cheapest things we could find at parked food trucks, drank the cheap wine we had transferred to a couple of colored canning jars, and we had a ball. We visited three or four National monuments, swooned over the Declaration of Independence, and imitated the monkeys at the National Zoo. Pete hung around with her roommate, but nothing came of it; she was too good to actually date a sailor. Too good for an enlisted sailor, that is, but I bet she'd jump the bones of a Naval officer in a heartbeat. Unfortunately, there was no jumping of anybody's bones on this trip.

Our final radar course was coming to an end, and two classmates talked Pete and me into renting a three-bedroom apartment with them on the beach at 63rd Street. One bedroom had a double bed, one had a single and another had two twins. That accommodated the four of us. After completing our training on our A6 jets, Pete and I were mustered into VA-85, Dan was sent to VA-35, and Kenny went to VA-75. Since the four of us were in three different squadrons, it was virtually impossible that all four of us were home at the same time. Only one A6 squadron was assigned to each aircraft carrier when it went to sea, so one or two of us were always gone. Anyone entertaining a lady got the double bed. The apartment had a washer and dryer, so we kept the linen clean, and a nice kitchen so we cooked at home.

The lure of the Dunbarton girls kept calling us. Dan had a car, and Pete and I borrowed it often. Dan never went to see the girls because he was very bashful and had trouble talking with girls. His dad was the postmaster of a small Pennsylvania town near Trenton, New Jersey. His mother died when he was a kid, and he had been raised like a monk. His home sat alongside the Delaware River, right where Washington crossed to attack the British near Trenton during the Revolution. The historical aspects of the scene were most impressive; crossing a semi-frozen, ice filled river in the middle of winter at first light was quite a feat by any standards. Dan was very quiet around people unless he had a few beers, at which time he revealed his keen wit. He was a hard-working, honest twenty-two-year-old Polish boy with a heart of gold who also smoked an occasional joint. He was a crazy adventurer who bugged me to death about riding motorcycles out to the Grand Canyon during one of our leave periods. I refused to put my young life in that much danger; I hadn't spent two hours total on "murder cycles" in my life, so he went alone. Obviously stoned, he wrote a terrible, but touching, poem called "Witty Smitty" one night in the middle of the desert. He returned in one piece.

Kenny grew up in New Jersey, and emotional problems surfaced after he joined the Navy, preventing him from finishing his four years. He got along well when he was not on a ship, but we were told by his squadron mates that he became a basket case at sea. That was a shame because he was a good, fun-loving, attractive kid who loved country music. His squadron buddies rented an apartment near us and had a record player and four Patsy Cline albums. Kenny would get wiped out and play those things all night long at the highest volume the old machine could produce. Thank God, he usually hung out with his shipmates from VA-75 when he was crazy. He never drank at our apartment. I don't know if drugs played a part in his demise, but I never saw him take any.

Many times, Kenny acted so weird I was afraid he'd try to kill himself. At night, five or six of us would go to the beach across the road from our apartments and body-surf in the waves at high tide. He would go out farther and farther until he found a sandbar and would stand out there in those thundering waves forty yards from shore in the deep darkness of an unlit beach, until we were all sick with worry. He would come in after a long while and fall exhausted on the sand and cry and cry. Eventually all his long and expensive training went down the drain because the Navy would not repost him to a permanent duty station on shore where he could have some stability. Kenny's poor mental health forced him to be examined by Navy doctors in Washington. He was pulled out of active duty in February of '69 in Fallon, NV while his squadron was practicing touch-and-goes and bombing runs for a deployment aboard the U.S.S. *Saratoga*. He was hospitalized for observation and evaluation, never to return to us. Pete and Dan and I continued renting our apartment with just the three of us; we never looked for another roommate.

On each of my next two trips to DC my old friend had a date, and I kept ending up with her same roommate. I didn't know whose idea it was, but I enjoyed it. The roommate's name was Jane, and I wrote to her about how nice her DC friends were. Virginia Beach and my squadron came up often, as did comments on items in the news about Vietnam and the war.

One Thursday after work, Pete met a girl at a single's club on the beach. He invited her to our apartment to meet the guys. My impression was she came to the beach on a cheap off-season vacation to shack up with someone. She thought she was pretty cool. Dan and I were sitting in the living room reading and listening to music on his old record player. The screen door opened, and Pete and his new friend Josie walked in. Quick introductions were made, there was a little small talk, she was shown around, and they left. He came back late that night. The next afternoon, a Friday, Pete was working late cannibalizing one of our planes when Josie opened the door and entered without knocking. She flopped onto our big armchair and said, "One of you guys get my luggage out of my car. Pete invited me to stay here this weekend."

I really didn't like the looks of her; she wasn't that attractive, and I certainly didn't like her attitude. She was crazy to demand something like that from us without a please or thank you or even a simple hello, so I just sat there and said, "Get your own damn luggage. Who the hell do you think you are?"

Josie was taken aback and said, "I'm a lawyer, and you guys are just sailors. You should see the men I date. I wouldn't even talk to you on the sidewalk back home. Go get my bags; they're in my trunk."

Josie was in reality a paralegal working in a law office in Levittown,

Pa. She was right though, thanks to that legal savvy she had developed copying documents, we were just plain ole common sailors, but so was Pete. There must have been something special about Peter Boy that she found appealing, or she wouldn't have talked to him either. I had seen him in the shower often enough to know what it was.

Dan got up from the sofa and walked outside. In a minute he was back, holding one of her large suitcases in each hand. I said I hoped she didn't expect us to do her laundry, too. Josie just looked at him and didn't say thank you or kiss my butt. She walked into the bedroom that held the double bed and said, "Somebody needs to get their shit out of here. Pete said we'll be using this room for a few days."

Dan had never slept or done anything else in that big bedroom, so I got up without any sarcastic or unpleasant remarks and moved my stuff out. I wondered what they would do in there for the next few days if it was beneath her dignity to talk to a common sailor. When Pete got home from work, he grabbed some clean clothes from the dresser in his regular room, and a beer and some chips and cigarettes from the living room cocktail table, and went straight into the big bedroom where Josie was waiting. With all their high-pitched squealing, it sounded like they were having a greased pig contest in there. Dan and I went to Burger King without even inviting them.

Thanksgiving of '68 came and went. I complained to Jane in a letter that I was a little depressed. I had been thinking of her, my folks back home, and all the things I wanted out of life. "I felt like my Thanksgiving had been bought at some bargain basement store. It was the wrong size, the wrong shape, and definitely the wrong color, unless you prefer blue holidays. We ate on base and had turkey and dressing and pumpkin pie. I hate to see sailors' families come to these dinners, because it ruins the atmosphere for me. I don't like to see civilians around the base on holidays, because it seems so bogus, so fake, and makes me sad. Pete and I went back to the apartment and watched a football game.

At halftime, the band crowded the field and played a lot of patriotic songs, which were sung by a large choir and the audience. I should have enjoyed it, but it made me start to think."

In my foul mood, I went on, "People sing songs and send pretty letters to the newspapers, but deep inside they really dislike servicemen and try to take advantage of them when possible. They sing, 'As He died to make men holy, Let us die to make men free…' but they tend to forget who's doing the dying and for whom. I realize that I'm safe and have little chance of getting injured in my branch of the service, but there are good American boys getting killed all over the world every day for our country, and yet everyone wants to protest and think they're so much better than those servicemen. Those people should thank God that they have someone to protect them."

I felt saddened and betrayed by the general public's negative reaction to the Vietnam War, but their displeasure was just beginning to spread. The next couple of years of lawlessness against our servicemen and policemen really took its toll on me. Pete took a different approach in expressing his disappointment with our fellow Americans. He had that mean streak when he got riled up. He would pick a fight with someone whenever he heard them talking trash about servicemen. It wasn't just because we were servicemen ourselves, he really thought America's servicemen and policemen were head and shoulders above the average American. He was sickened when he saw a line of protesters throwing rocks or buckets of water at cops. He railed at the occasional newspaper he read, or the news report he inadvertently saw as he walked past a television, as they described the burning buildings in Detroit during Devil's Night, knowing no one would ever be punished for the wanton destruction.

Pete hated to see some smart-ass college kid burn an American flag or join some protest against the Vietnam War. At a friend's apartment one night, he was enraged when he saw some Vietnam vets throw their medals away during a nationally televised protest. I had to get up from the sofa and walk over and change the station before he kicked the screen in; I'd seen him do it before. I almost didn't make it in time; the channel knob was missing, and I had to find the pliers to turn the dial. He wanted to strangle the excuse-making politicians who played to their constituents to garner votes at the expense of our country. He was disgusted at the sorry people who were beginning to make up America, especially all the paid "outside Communist protesters," as he called them. He believed that anyone who didn't love America should leave it. He always said, "Delta is ready when you are."

Our work sheltered us from the constant protests shown on the daily TV news and blared out on the radio; we had no such outside distractions in the hangar. And to insulate us further from the divisive political unrest which made us sick, we didn't own a television. We didn't watch sitcoms, the news, sports or politics, and we didn't depend on the radio for any news. We got our news via rumor and scuttlebutt. Nothing about that has changed in the last fifty years for young people, except the quality of the rumors has decreased considerably over the years in direct proportion to the increase in its quantity. We were young, single, working men, living in a dream world of jet aircraft and enormous ships, and we didn't give a damn about anything except naval events and girls. Pete and I were isolated from the world and were reasonably happy, as were most of our squadron mates, many of whom had volunteered to serve.

Our new happy home was soon to be with the Atkron 85 Black Falcons, as known in military circles, or Attack Squadron 85. Feeling a little lonely and missing Jane, I wrote her a short letter, "I'm training with VA-42 at present, but VA-85 will be home from sea soon, and I'll join them here. I've heard the whole squadron gets off from December 12th to January 5th, and if so, then I'll go home."

I then brought up the real reason I said I was free during that upcoming three weeks over the holidays. "But if I don't go home, I was wondering, if I send you round trip plane tickets from DC to Norfolk and make arrangements for a visit, would you consider joining me here for a few days? It does sound like a disreputable proposition, but I'd really love to see you." I had no actual intentions of going home, but in my defense, I was weary and light-headed while writing the letter.

"I've worked 37 hours straight, except for 3 hours sleep and 2 hours for food." That sounded quite heroic, except there I was, still at work, sitting on my butt, writing a letter to my girlfriend. I hoped she would come even though her family expected her at home, and she had no legitimate excuse for not being there. But I had an apartment at the beach, and I dreamed she would want to join me.

We enjoyed a week together, then Jane headed to Pittsburgh to see her folks, and I stayed in Virginia Beach. Pete knew my holiday plans didn't include him that year and went to Augusta. Dan spent the holidays in Pennsylvania. It was almost a week before Christmas.

"Yesterday an F4J Phantom was lost at sea about 40 miles from here. It was in Squadron VF-102, and both the pilot and the bombardier-navigator were lost. It'll really be a terrible Christmas for the poor families of those two guys. But to everyone who reads about it in the newspaper it is just one of those things, and to me it is just news, and to you it really should mean nothing at all. But life is like that, isn't it? I mean, if something doesn't directly affect you, then it has no real meaning for you, and you don't care about it. But one of these days, no one is going to care about anything, and we'll be in a helluva mess. Maybe you and I should start caring a little more ourselves."

I added what Pete had said to me, tongue in cheek, one morning when he came in from the convenience store, "Ahh, this good Christmas spirit makes my little ole heart sing. A woman actually spoke to me this morning; she must not have realized I was a sailor. Everyone on the block is moving out now that someone told them we are sailors, must be something about undesirable neighbors." Pete was calling us, and no one else, the undesirable neighbors, but he was just kidding; we got along very well with all our neighbors, in all their many colors. And for the record, Negro was the only N word we ever used, but black was in vogue and was used almost always by us. African American was way too politically correct at that time, and the beautiful thing about black was it has only one syllable. We certainly respected and befriended our black brothers in our squadron and on the ship, at least I did.

Occasionally, I went visiting with a fine black shipmate, usually to Norfolk. It was pretty wide-open and was loaded with sisters. The father of one girl my friend Chuck was dating at that time had a hard time believing a tall good-looking black kid from Oakland and a skinny white boy obviously from the Deep South were friends, but we were. Pete never hooked up with Chuck because he thought associating socially with blacks would make people think less of him. His real problem with Chuck was most apparent when at sea, "Man, if you invite Chuck to go ashore with us, we'll never pick up any girls."

In reality, that was just Pete. I had seen no racial disharmony in the Navy since peace and quiet had been restored in Memphis after Dr. King's death eight months earlier. But I wasn't a good judge of race relations, because my good friend Chuck was the only black man I knew personally during my four Navy years; soul brothers were just not there in naval avionics. Besides, he couldn't have hurt anything anyway; we never picked up any girls, with or without being accompanied by a black sailor. Jane and I always welcomed him at our apartment in base housing; he ate with us often, but we drew the line when he brought a love-interest with him.

He'd introduce her to us, and we'd have a couple of drinks. We'd talk a little, and then he'd pull me aside and say, "Hey, man, can I take her upstairs?"

"No, sir," I'd say, "not in my bedroom, especially not in broad daylight; go get a motel room."

"Smitty, you know I don't have enough money for that," and he'd roll his eyes round and round and stick out his lips.

"Besides, man, we don't want a bunch of kinky hairs all over our bed," I'd say, and we'd all laugh and laugh, including the girl, one of whom once replied, "It's okay; we won't pull the covers down."

Before our Mediterranean cruise, Chuck gave me a fine autumn jacket with a plaid inner lining for my birthday. I wore it for nearly ten years until I got too big for it. I gave it to the Salvation Army, and I thought of him.

Dealing with the local white surfer population was a different story of disharmony. "I've got a great story for you," I wrote Jane. "The guys were talking about it in the hangar, and they swore it's true, but that doesn't make it so, does it? Gary Osborne was walking to his apartment late Friday night and had to pass by the Tiki, a local beer joint where about twenty long-haired pseudo-surfers were standing out front. According to Gary, one of them said something meant to piss him off, so he stopped and asked the guy what he said. The guy refused to answer, so Gary turned and continued down the street toward his apartment. Out of the dark, a baseball bat slammed against his back, knocking him down, where he was kicked and beaten. He slithered between some legs and started running, pulling out a switchblade knife as he ran. He was chased until he turned around and slashed two of his attackers with the knife, drawing blood and momentarily stopping them. Gary reached his apartment and ran up the stairs, his knife and keys falling through the steps to the ground. The door was locked, so he yelled to his roommate to open it and get the butcher knife. His roommate Bill was panic-stricken and ran in a little circle before letting him jump inside.

"The surfer boys arrived and were banging on the door. One guy hammered on the window next to the door with a crowbar or something until it caved in. A hand reached through the hole, trying to unlock the door and the safety latch. Gary grabbed the butcher knife and started stabbing, cutting the surfer's hand so badly it needed stitches.

"The surfers charged the flimsy door and knocked it down. Gary jumped through a bedroom window. He landed on a truck outside, flipped to the ground and took off for the Hamburger Haven next door. In the meantime, Bill had also jumped and landed somewhat less gracefully, breaking his left arm and cracking a vertebra in the small of his back. Meanwhile, Gary was holding everyone in the Hamburger Haven hostage using two large carving knives he found in the kitchen. Bill crawled over to the Rathskellar, but they wouldn't let him in to call the police. They didn't want any trouble, they said, so they called the police themselves. Gary had called the Shore Patrol from the Haven and said he wouldn't let anyone in or out until they came.

"A friend of Gary's picked a poor moment to visit and was caught by the gang as it poured out into the street trying to find Gary. The friend was beaten, and all the windows in his car were broken. The cops came, and Gary released everyone in the hamburger joint. He went to the patrol car and explained what happened. The cops then listened to a version of events given by one of the long-hairs and decided to take Gary to jail. It was an easy decision; it was a sailor versus a local surfer boy. They took Bill to the Portsmouth hospital and let everyone else go. One of the surfers who was slashed on his shoulder went to the emergency room where he was interviewed by the police, and he told them what really happened. Gary was soon out of jail, and the police asked him if he wanted to press charges. He said, 'Hell, yes! I'm getting a lawyer.' I heard Bill is in traction; they'll operate on his back soon, and after he recuperates, he'll be out of the Navy on complete disability."

I enclosed a clipping from the *Virginian Pilot* dated December 16, 1968, that reported the return of a great warship and VA-85, my new squadron, my new friends, my new family, for the next two and a half years:

"U.S.S. *America* due Today after 8 Months.

"Norfolk—The Norfolk-based carrier *America* returns home today after an 8-month deployment to Vietnam.

"The 77,000-ton ship, which left here April 10, will be welcomed home with military and civic ceremonies at Naval Air Station Pier 12 at about 1:45 p.m. Most of her air group flew in Sunday. Government and city officials will greet *America*, as well as the families of the 5,000 crewmen aboard.

"Fireboats spraying colored water will escort the ship from Fort Wool to the pier, where the Washington High School Band will offer a musical greeting.

"*America* arrived on 'Yankee Station' in the Gulf of Tonkin May 30. Air Wing Six, based at the Oceana Naval Air Station, flew their first strike missions the next day.

"Before they left for home, planes from *America* dropped more than 18,000 tons of ordnance on the North Vietnamese in 112 days. They destroyed 1,114 trucks and 454 waterborne craft and interdicted enemy supply lines.

"Ten men from the air wing were lost in combat—one killed and nine missing in action.

"Air wing pilots also flew spotter missions for gunners of ships operating with the coastal forces, including destroyers from the Royal Australian Navy.

"*America*'s pilots flew 11,000 combat and combat-support sorties during the deployment."

The day after Christmas, I wrote Jane, "Nothing new has been happening around here, except Dan came back from visiting his dad. The lady from the apartment where we borrowed the salt and pepper brought us a completely cooked turkey stuffed with dressing and stayed for a couple of drinks. We ate a lot of the turkey and talked awhile. She's great, and my Christmas did have a turkey in it. I only missed you to make it perfect." Her act of kindness made me take back all the negative things I had said about the civilians of Virginia Beach. Pete and I often played with her ten-year-old son after that; he had been a stranger before then. I never asked her where his father was.

6. SAN DIEGO SOFTBALLS

Our new squadron had finally settled in since returning in December from Vietnam, and Pete and I officially joined them on January 5, 1969. Seven sweet months of fun and sun in Oceana, and then VA-85 was to begin more Vietnam bombing runs in August. Pete and I were too ignorant to be afraid of heading to the Gulf of Tonkin, sitting in the middle of Yankee Station on a floating ammo dump filled with jet fuel, rockets and bombs that could explode at any moment. Neither of us were familiar with the U.S.S. *Forrestal* fire.

"Relax, man," Pete said, "nothing bad's gonna happen. These carriers are the safest ships afloat."

But, in July of '67, just seventeen months earlier, during combat operations on Yankee Station, an unholy conflagration engulfed the *Forrestal*, and 134 men were killed and another 161 were severely injured. To minimize the effects of similar occurrences in the future, the Navy set up firefighting schools for all Navy and Marine personnel. The east coast school was in Mayport, near Jacksonville, and I was chosen to fly down there for a few days. My squadron kept sending me to places like that; they either thought I was going to make the Navy my career, or they wanted to get rid of me for a while. It was probably the latter.

Attendees at the firefighting schools were run through burning towers and smoke-filled rooms to their eventual safety. We stood for a few minutes in a sealed room similar to a Nazi concentration camp gas chamber, nervously chatting about our adventure, waiting for our next lesson, the gas masks we were given in our hands. An instructor pulled a silver tear gas canister from his pack, pulled its lanyard and threw it in the middle of us. As he ran from the room, he screamed at us to stand away from the smoking tube or it would catch our clothes on fire. It flamed and sizzled and hissed; its wispy white gas flowed from its top, filling the room. We choked and coughed and cried as the escaping chemicals entered our noses and eyes, and our mouths and lungs. The gas irritated our mucous membranes and forced snot from our noses and washed our eyes with tears, until we were blinded and pitiful. The petty officer waited another minute and jumped in the room and yelled for us to put on our masks. We had suffered until we could stand it no longer and clumsily fumbled, sick and blinded, to strap them on. Thirty seconds after donning our masks and completely sealing them from outside air, we could breathe and see again, thus proving to us that the masks worked and convincing us to put them on during a fire. The more extensive training given to the men who would become firefighters paid great dividends many times in the Navy's future.

On January 14, 1969, nine days after Pete said nothing bad could happen to us, the U.S.S. *Enterprise*, nuclear carrier CVAN-65, sailing west to the Gulf of Tonkin, was crippled by disaster off the coast of Pearl Harbor. During flight operations with fully armed missiles, the exhaust of an F4J fighter detonated the four Zuni rockets in a pod mounted on the jet behind it. They flew down the flight deck, setting many more jets loaded with missiles and jet fuel ablaze. The ship's firefighters and on-deck crews bravely fought the burning fuel and exploding missiles and saved the ship from complete destruction. When all was secure and the order to stand down from General Quarters was given, 28 men had been killed, 343 were injured and 15 jets had been totally destroyed. Because the firefighters had lived up to their great training and had handled their jobs so well, the *Enterprise* was quickly repaired at Pearl Harbor, and in March of '69, she returned to Yankee Station.

Pete misspoke about nothing bad happening, but we would have been well compensated for any trauma we might have endured at the hands of the Navy. I got my 1968 W-2 and was amazed I had earned a whopping $1,677 for being a full-time sailor for the entire year. Man, that's $139 a month, every month, almost $35 a week. Wow! Of course, I did receive free room and board, and medical care was available had I needed any, which I didn't. When I joined the Navy, I was told death benefits were $10,000, but nobody told me the benefits came from a life insurance policy whose premiums were withheld from my pay. All my uniforms, shoes, and work boots were paid by me. A monthly draft from my pay went to the Citizens & Southern Bank in Augusta to repay my $10,000 college loan. I bought my smokes each payday and drank the rest of my take-home pay on liberty.

I wrote Jane a late-night letter, "Today was beautiful with clear skies, gentle breezes, and warm temperatures. The runway at the base was classic tonight with planes, and noises, and lights. I enjoy walking the flight line while the planes are flying and feel the mist of diesel and jet fuel on my face. It makes me feel like I'm doing something after all. When the jets hit after-burner on the flight line at night, the colors are magnificent.

"Kenny leaves Monday for Fallon, NV on a detachment and comes back in 6-7 weeks. I leave in 8 weeks on April 19th and come back on May 3rd from Fallon, so I won't see him for a while. I'll be going to San Diego from June 23rd—July 15th for carrier qualifications. We are leaving for Vietnam on August 20th. The squadron will not come back from San Diego between the 15th of July and the 20th of August, but I can take personal leave and fly home on a Navy plane. So, we have to plan something a little later for our rendezvous. This Navy crap may be a pisser for our personal life, but we must remember that someone has to do it."

As time got closer to our deployment date, we had to finalize our business obligations. Two other sailors took over our interests in the apartment on 63rd Street, and Pete and I went back to living in the barracks. That was fine with me, but I wanted to see Jane in a nice quiet setting one last time before I went to sea. I had leave coming after our sea qualifications in San Diego, so I called Dan to see if Jane and I could stay with him to be near the beach. I knew he wouldn't make me pay anything. He said come on over; his new roommates had gone to some three-week specialized avionics school, and there was plenty of room.

"What about Kenny. Will we bother him?" I asked.

"Oh, I haven't told you. They pulled him out of Fallon and sent him to the shrinks in Washington. He just can't hack it; he's sick. I heard he'll never be back."

"Did you get to say goodbye? Did you see him?" I asked.

"No. A Chief came to the shop to check me out and asked if I would pack his personal things at the apartment. Then two guys came over and took it away. I have no idea where he is."

"Man, that's tough. I don't know what to say. He was a good guy," I added.

"That's all you need to say."

Jane and I enjoyed our short time together at Dan's. During the night of July 20th, the three of us watched "one giant leap for mankind" on a small black-and-white television he had bought on lay-a-way at Sears just for that event. If you ever wondered how much money ordinary sailors had, think about being forced to buy the smallest, cheapest black-and-white television set Sears sold on lay-a-way. After the giant leap for mankind, Jane and I walked down our one-lane, oyster-shell-and-asphalt street to the beach hand-in-hand in the dark and strolled along its deserted sand for an hour or two, feeling like we, too, were on the moon with Neil Armstrong.

A week later our aviators were sent one last time to combat weapons training school at NAS Fallon, and only our most experienced enlisted crewmen had been assigned to assist them. We rookie crewmen would get our real training in the field, or rather on the sea.

Every piece of squadron equipment was loaded on huge cargo planes and flown to NAS North Island in San Diego. The gear was then loaded aboard the U.S.S. *Constellation*, CVA-64. During the days we were preparing for sea operations, our aviators had flown our A6As to NAS Miramar for additional TOPGUN training maneuvers at the Navy Fighter Weapons School which had just been started that March of '69, four months earlier. Miramar was called "Fightertown USA" by all the naval aviation boys. Rumor had it that our pilots and our jets were so superior to the other squadrons that our boys couldn't even see who was second. Later, when the *Constellation* was prepped, packed, and sea-ready, she would pull out of North Island, ease into the Pacific and head for Hawaii. The squadron aviators who had been at Miramar would join us along the way with their first shipboard landings of the cruise.

The Navy flew our squadron's enlisted men on chartered jets to NAS North Island on the Coronado peninsula on San Diego Bay. We grabbed our gear and jumped aboard one of the many cattle cars taking us to the ship. This was the first introduction to an aircraft carrier for many of us. We were awed by her size, awed in the true sense of the word: she scared the hell out of us. She was a thousand and eighty-eight feet long and two hundred and eighty-two feet wide, and she could make thirty-four knots, about thirty-six miles per hour. In plain English, that made her three hundred and sixty-three yards long and ninety yards wide, and at thirty-four knots, she'd sail eight hundred and sixty-four miles in one day. The tower on the side of her flight deck was fifty yards high. Pete and I loved her from the start.

"Look at that magnificent son-of-a-bitch," I said. "Look at the size of that thing. God, she's beautiful."

While the traditional black-shoe seamen, collectively known as the ship's company, fitted the ship for her deployment to the Pacific, we aviation boys had a lot of free time on our hands and no place to go. Not knowing anyone in San Diego, we naturally just walked around and patronized bars that catered to enlisted men. I wished Jane had been there with me instead of Pete. No offense to him, but there were more things I could do with her. Sitting at a window in a burger shack eating a double cheeseburger, I spied my friend John Hayden strolling by, all alone.

"John...John," I yelled, as I ran out the door, "whatcha doing here, man? Last time I saw you was in boot camp."

"Hey, Smitty, is that you? What you doing, buddy? My God, you look good," he said, shaking hands and giving me a little hug.

"Come inside," I said, "talk to me. Come meet my friend. Let me buy you a beer."

I introduced the guys, and we ordered another beer. John was a great guy, and it was good to see him. I asked was he shipping out soon, or was he stationed in San Diego.

"Man, I'm going to Nam," he said.

"Really? So are we. What ship you on?" I asked, thinking it would be great if we were all on the *Connie*, as we called her.

"No ship, man; I volunteered as a door-gunner on a helicopter. In country. That's the God's truth."

"You're lying! You volunteered? Are you crazy?" I caught myself and stopped talking like that because I was afraid he might actually be crazy, or on drugs, and I didn't want to upset him. "What happened? You're in the Navy, not the Army."

"That's okay; they'll take anybody. I saw a poster in the chow hall one day. They need volunteers real bad and gave me a signing bonus and a bunch of other crap. I'm leaving at three and still haven't packed my duffel bag. I gotta go as soon as I finish this beer."

"Where you been training? What weapons you gonna get? Where's your base?" I asked.

"Well, my plane takes me to Cam Ranh Bay, and then I'll be with the 54th General Command wherever they send me."

"So...what's your weapon and your copter?" Pete asked again.

"Yeah, that's gonna be fun. I'm going to fire an M60 machine gun from a UH1 Huey about a hundred feet off the ground going eighty or ninety miles an hour at any frigging thing that moves."

"That's awesome, man. How much bonus did you get? Is it worth the risk? Where'd you go to weapons school?" Pete said.

"I'm not supposed to tell anybody what my bonus was, but who cares now? The Chief who signed me up gave me a thousand in cash, not a check or a bank draft, but cash, out of his own pocket. Ten brand-new hundred dollar bills. Can you believe that crap? I'm not sure that was even legal."

"So where was your weapons school?"

"Man, you keep on asking me that...I hate to talk about it because it's the only thing that scares me. I've never even seen an M60 machine gun. I've seen some pictures, sure, but I've never seen a real one. Can you believe that? I'm going to war in a helicopter, firing some frigging weapon I've never touched before at some stinking little yellow gooks running through the jungle in their black pajamas."

"Je-sus," I said, "you better be careful, man. Surely they'll send you to a school over there, but the whole thing sucks."

"Tell me about it," John said. "Oh, by the way, did I tell you that the average life expectancy of a door gunner in Nam is fifteen to thirty days?"

We looked at our beers and didn't say a word. John glanced at his new Rolex watch and chugged his beer, and we shook hands goodbye. "Good luck to you, ole buddy," I said.

"Thanks, man, it's been real. Hey, I got these brews. My signing money," he said as he dropped a twenty on the table and walked out the door.

The next day, Pete and I were downtown in a bar on a side street, sipping a beer, and talking about John. The curtains were slightly closed, and the room was dimly lit. Business was kinda slow; only two other customers were there, possibly because it was the middle of the day. Music was playing on the jukebox as we talked at the edge of the pole-dancer's platform. The place served Heineken on draft, and that's what we wanted. The bartender brought a large pitcher, its top circled in ice, and slid it in front of me. He returned with two frosted mugs, some unshelled peanuts and an ashtray. I poured each of us a beer and lit a cigarette.

We sat on cushioned stools along the long empty stripper's runway, minding our business, sipping beer so cold the heavy mug had an inverted cone of ice in its center. One of the customers, a scantily-dressed woman, got up and walked over to us, apparently to chat. I had barely noticed her when we came in, but now that she was on her feet, I saw she wasn't a customer at all. She was dressed like a go-go girl with tassels hanging from the front of her canary-yellow halter top and matching sexy miniskirt. The tassels reminded me of the old Roy Rogers cowboy suit Santa brought me when I was six. The skinny strips hanging from the bottom of her bikini top mercifully covered her stomach that was ribboned by stretch-marks from a pregnancy. Her large firm breasts, the size of softballs, strained against the confines of her top, trying to free themselves. The money she had invested in them was well spent; they were great assets for her job. Her peroxided blond hair, gathered into two long pigtails in American Indian style, had frizzy split ends and ugly dark roots. Her face was heavily made-up, and I hated the thick black mascara underneath her eyes.

The barfly stopped at the barstool two places down from me. She had a mixed drink in one hand and a cigarette in the other. She took a sip of her drink through a tiny red cocktail straw and then inhaled deeply from her cigarette. She was obviously health conscious because she used a little plastic filter on the end of her cigarette to prevent her from getting lung or throat cancer. She exhaled slowly, blowing a thick stream of smoke a couple of feet above her, and began dancing with herself while the Supremes sang *You Can't Hurry Love* on the jukebox. All the while, with one arm raised above her head and twirling like a ballerina, she stared directly at me. I thought it was the most embarrassing thing I'd ever seen…until she climbed up on the platform. The music changed to *Born to be Wild* by Steppenwolf, and the woman pranced toward me in her bare feet, running her fingers through her hair and smiling at me like she was a lion and I was a helpless gazelle drinking at the edge of an African lake. She stopped directly in front of me. She began to gyrate like a crazed Arabian belly dancer. She reminded me of a thirty-five-year-old woman playing with an invisible hula hoop in slow motion on a theatre stage.

My biggest objection to the private showing we were honored to receive was that the woman was standing directly above my beer. From my focal point, just by looking up and without leaning forward, I saw through her thin canary-yellow panties that she was indeed a lady. But I couldn't decide whether she always acted like a lady so I was concerned about any health conditions which she may have possessed, and I put my hand over our pitcher of beer to save it from any fallout.

Having been a smart-ass in my earlier youthful days, I saw no reason to show any restraint at this time. I looked up and told the woman, "You're polluting my beer."

Steppenwolf had finished his song just before I said that, and the woman heard my comment clearly. She stopped her belly dance and glared at me. Talking in a crude, fake Filipino accent she'd undoubtedly picked up from some other sophisticated naval personnel, she said, "What's the matter, sailor, you no horny?"

No one had ever said that to me before, so I was caught a little off guard, but I had heard some of our guys laughing about it from their last cruise. Of course, I had to have the last word, "Get out of my face, lady. Leave us alone."

I was the one who had unfortunately started this quarrel with my comment about her polluting our beer, and her response about being horny was probably made facetiously and didn't mean any disrespect to me. I overreacted, spoke too soon and too loud, and in the stillness of the bar between songs, my voice carried to the other customer still sitting at his table. A large middle-aged man stood up and headed toward me. I realized then that he was no more a customer than she was; he was her husband, her protector, the club bouncer.

The man was huge; he was twice my size, and I was scared. As he walked across the room, I thought, *what is this son-of-a-bitch going to do?* He was the bouncer and was paid to kick trouble-makers out of the club through the use of brute force and violence. He probably had a knife or a pistol on him, and I had nothing. He took a final step and grabbed me by the neck with his left hand. He raised his right hand to slam his fist into my face.

Suddenly, before he could extend his fist, Pete delivered a crushing blow to the side of the man's head, sending blood everywhere. The man released my neck, paused a second and staggered backward. Pete's mug was cracked in half, and he held a part of it by the handle. Blood dripped from the jagged edges and onto the barroom floor. The woman screamed as her husband hit the floor, thankfully face up. He was unconscious and bled from a large gash above his right ear. The strength of Pete's blow peeled a flap of skin from the man's scalp and exposed the white bone of his skull above his temple. Pete dropped the mug and said, "Let's get the hell out of here."

Pete and I looked at each other and didn't say another word, not to her, not to each other. We froze at the enormity of what he had done.

The man's gash was deadly serious, but Pete had only defended me. A loud metallic noise, like someone dropping a pistol, clanged in the back behind the door to the kitchen, releasing us from our astonishment, and we grabbed our hats and ran out the front. I was so excited; this was what being in the Navy was all about, and we hadn't even left the Mainland. All the tall tales we had been told were true.

Pete and I ducked into another club five or six blocks down the street and rinsed off our few splatters of blood. We spent the rest of the afternoon drinking, trying to come down from the emotional high of Pete's assault on the bouncer. We paced ourselves, but time and volume eventually took control. The more I drank, the more I wondered about that woman's beautiful softballs; they kept getting bigger and better. Thankfully, I was almost penniless, and it was late. Besides, we could never return to that club, and I would never indulge in that sort of thing anyway. We reported to the ship, so we wouldn't be written up in the morning. When we got back to North Island, we went straight to our bunks.

Pete and I were ready to ship out. We had seen enough bars in San Diego. We had even gone to the world-famous San Diego Zoo. It certainly lived up to its great reputation. The best lesson I learned there was to never piss off an alpaca; that son-of-a-gun can spit a wad of mucus about thirty feet. San Diego was a beautiful city, and the weather was excellent, but we were ready to see more of the world. We were itchy, we were bored, and a bored young man of any age can easily get into trouble.

One place on the west coast that Pete and I were eager to see was Tijuana. We knew it as Tia Juana, its original name, and we knew it was wild, a place where anything goes. Pete reminded me to get a pass to cross the border into Mexico. Passes were essential to any sailor leaving the country and wanting to get back in some day. Pete got his, but I forgot to ask the Chief for mine. When I saw Pete getting ready to shower that Saturday morning, I knew I had screwed up. I hurried to our AQ shop where we worked, but the Chief had gone ashore. I was mad as hell, not at him, but at me. I was looking forward to the trip, but I had been on the hangar deck all day Friday cannibalizing one of our planes and had simply forgotten all about it. It was my fault, but I hated to miss a chance to see Tia Juana with my buddy.

I knew it was wrong, but I got ready and walked off the ship with Pete. Buses always picked up sailors around the ship, taking them many different places, so we took one to the U.S. border. Getting into Mexico was easy; you just strolled across the bridge spanning the Tia Juana River, walking through the caged tunnel beside the roadway. There were no security checks leading into Mexico, no one cared who went in, but getting back to the U.S. required going through check posts.

We had grabbed one of our shop friends as we left the ship. Gabriel's mother was French and his father was Basque, so he spoke fluent French and Spanish. He was from Miami and was a great kid and a great asset in Mexico. We ate tacos from street vendors, we drank all the local beers, and we spent a couple of hours at the bullfights. We listened to the old men playing Mexican songs on the sidewalks and drank some more. We posed with beer bottles and giant sombreros. Our picture was taken as we sat like three fools on Mexican burros on a street corner. When the old men who owned the burros laughed with us, their large white teeth sparkled in the dazzling sunlight. Their sun-dried skin was dark and leathery like Egyptian mummies; they looked like extras in the Bogie movie about the Sierra Madres. I gave one of them a quarter to say, "We don't need no stinking badges." We roared with laughter; we may have been drunk.

As night fell, Gabe got us into a restaurant open only to Mexicans. The man at the door told all Gringos they were closed, and we were no exception. In old Spanish learned from his father, Gabe told the man we only wanted something authentic to eat, and we would behave. He looked at Gabe and saw his deep olive skin and dark hair and features, smiled and opened the door. The three of us went in, sat at the counter amongst the Mexicans and ate a great meal. We watched our manners and quietly talked as they kept putting more food before us right off the grill, very much like our fish fry at Blue Lake. They wouldn't let us pay for a thing.

We went into a nightclub for one last drink. There was a floorshow starting in a half hour, so Pete said, "Let's see it."

We had another drink and another. By the time the show started, we were two sheets to the wind. The star of the show, a lady whose waist was a little larger than mine, began gyrating and side-stepping down the runway, heading straight for our table. Since we were the only Americans in the place, it was easy to see how she had made her choice. She stopped and twirled, and then began to dance the Flamenco to the music coming over the loudspeakers. It was very classy. She danced and stomped her feet; her Flamenco shoes looked like some old black work boots with their toes cut out and steel taps nailed to their soles. She flared her black-and-white checkered cape, possibly purchased during a Baja endurance race, and smiled with dark yellow, cigarette-stained teeth protruding from her garish blood red lips. Her complexion was reminiscent of the 'before' photo in a dermatologist commercial. She grabbed Pete by the arm and began pulling him from his seat. He apparently didn't care to dance, so he put up a good fight to prevent it. If it hadn't been for my help, he would still be sitting there.

In any event, the two of them danced over to the center of the stage, as the lady waltzed around Pete a dozen times, smiling and waving that stupid cape. Finally, the sound system began playing *Here Comes the Bride*, and someone threw her a bouquet of flowers. Pete was in a drunken daze as he was being led down the aisle to the preacher awaiting the dear couple at the runway's end. After the brief, but lovely, service, the lady turned to the audience and bowed, and everyone laughed and applauded. She turned to Pete who was almost too far gone to even notice and puckered up. She leaned into his face and, being nice, he gave her a big old smack in the general area of her teeth and lips. Everybody roared. I didn't know what was so funny until the lady made one last twirl and went down on one knee. She reached up and grabbed her hair and threw off her wig. The lady was a man, a fortyish, stocky, transvestite, drag queen who was totally bald and quite ugly. Pete's face turned blood red; he wasn't too drunk to realize he had just been had. He was humiliated by the crowd's laughter, and I knew what was coming. Gabe and I jumped up and ran to him onstage which increased the crowd's amusement even more. We grabbed him and hauled him away before he could kill someone or something.

"Let's get out of here, guys, back to the ship. The last thing we want is for someone to get arrested," I yelled.

The time had certainly come to head back to the ship. We took a cab to the border patrol checkpoint, and the security guards, all Mexican police, interrogated us. When they asked us for our military IDs, I pulled mine from my wallet. When they asked me for my pass, I fumbled around like I couldn't find it. They knew, and I knew, I didn't have a pass, but I didn't know which lie to tell. Pete and Gabe were motioned on ahead, back to the States. Pete was going to give them some crap, but Gabe thankfully pulled him away. They looked at me and walked away; they couldn't help me, and I settled into my cell. I sat alone in the Tia Juana jail for four hours while Chief Cranston processed the paperwork to free me and then drove a Navy vehicle from NAS North Island to the border to get me. There was nothing I could do but apologize; I had gotten him out of his rack on the ship to take me home. I told him it wouldn't happen again. And that was true, except for that time in Subic Bay. And that other time in Malta.

As punishment for my little Tia Juana indiscretion, I couldn't go ashore again until the ship was given liberty in Hawaii. Fair enough, no more San Diego, I wasn't going back anyway; I didn't have any money.

7. BURIAL AT SEA

The nonworking enlisted men of the ship's crew and the aviation squadrons gathered on the flight deck, lining the outside edge of our mighty warship in our clean, pressed white uniforms, as we slid through the smooth sparkling blue waters of Pearl Harbor. We occasionally lined the deck at some of the special ports we entered or left to show respect for that city or its country, but we didn't have the time or the energy to do that when we came off the line from combat maneuvers. All of us were especially proud to be standing at attention in our dress whites as we cruised past the U.S.S. *Arizona* and her memorial. Only during the solemn passing of the *Arizona* is there absolute silence on ships to show their respect for the thousand or so men still inside her, the men who died during the Japanese attack. Normally, the ship's electronic bells loudly tolled dong, dong, dong and her whistles shrilly piped as we passed in review, saying hello or goodbye all around the world and announcing that another great American warship was on duty. But that great hullabaloo was really produced just to spread the word from the United States Navy, 'Look at us, you nations of the world; look at our mighty power, we can destroy you, if you make us.' Not, if we want to, but only if you make us.

The harbor-master and several tugboats had come out to guide the *Connie* into Pearl when we arrived early that morning. Naval tugboats used to maneuver aircraft carriers weren't like the ones commonly found on the Ohio or the Mississippi Rivers. Those massive Pearl tugs nudged us into place with their tremendously powerful engines, and we were going to stay there for a while. Once we were tied up to our pier at the Naval Station Pearl Harbor, all the ship's company in their dress whites were allowed to go ashore. Pete and I walked the plank and stepped onto the glorious ground of the most beautiful state in the nation. We were joined by two squadron friends as we rented a car and drove counter-clockwise around the island of Oahu from Honolulu to the main highway's end at the western edge of the North Shore. We left the North Shore after sitting on the beach at Pipeline and wading in its awe-inspiring surf. We took the smaller road running north to south toward Pearl Harbor and drove through the pineapple fields of the Dole Plantation which seemed to take up most of the center of the island. The biggest shock to us during the entire day occurred when we stopped for gas; we paid thirty-five cents for a gallon back home and a dollar and twenty-five cents there. Our day's greatest pleasure was eating the finest mahi-mahi sandwich known to man at an old white food shack along the road halfway between Diamond Head and the North Shore.

We saw everything, ate almost everything, and drank almost nothing until we crashed bone-tired on our sacks back on the ship. My sack was a clean, but gray and dingy-looking, pin-striped mattress six-foot four-inches long, two-foot six-inches wide, and two-inches thick. Pete said his looked like something from a concentration camp and felt half-filled with lumps of cloth left over from World War II. Some sailors called them "fart sacks." These sacks laid inside metal slabs and together made up our beds, or racks. Rack was the perfect name for them, because they looked like the long rectangular baking sheets used to hold fresh pastries straight from an oven. Underneath the top surface where the mattress lay, there was a six inches deep storage space which flowed undivided across its entire six feet. The rack was hinged at the back, so sailors opened it from its long front while standing in the space between the three stacked racks in front of them with the other three racks behind them.

Between the stacked racks and at the end of each sleeping berth was a seven-foot tall locker, reminiscent of the lockers in a high school gym. In them, each sailor had a lockable two-foot metal cube to store the items that were valuable or wouldn't fit under his sack. The rack's storage space and the little locker held a sailor's uniforms, shirts, shoes, underwear and souvenirs. Enlisted men weren't allowed to wear civilian clothes during *Connie*'s West Pac cruise, so they didn't pack any.

Sailors hung their peacoat and work jacket in a communal locker at the end of their sleeping compartment, a giant bedroom holding ninety men; so crowded that sailors could see only ten feet in any direction. Dirty clothes hung from a ditty-bag at the end of each rack. The best rack was the middle of the three; the worst was the bottom, the one I got. The bottom sucked because every time the man in the top rack wanted to get up or down, he stepped on the metal edge of the bottom rack. That wasn't too bad, but we all worked at different times and had to go on watch often, so sailors went up and down all day and night. Each rack had a light like a coal miner's helmet recessed into the wall just above and behind a sailor's head. On the ceiling at different places in the giant sleeping compartment, uncovered lights burned brightly twenty-four hours a day, enabling sailors to see and to discourage any sexual activities, or thievery, or mayhem in the room. Each rack had a round, unfiltered, protruding vent on its side. By rotating the vent, the volume was adjusted to high or low, but there was no temperature control. Racks got cold air when the Captain said it was air conditioning season and hot air when he declared heating season.

Each sleeping compartment had an older sailor constantly on guard. On payday, our jovial guy yelled at the top of his black lungs,

"They's paying, they's paying; the bosses are paying." Everyone hopped up, got dressed and hurried to the hangar deck every two weeks, stood in line for a half hour and was paid in cold, hard cash. Each sailor signed a receipt on the paymaster's log for his, and only his, money.

An aircraft carrier at sea made its own fresh water, a slow and energy-consuming process. Most of it was for cooking, drinking and washing pots and trays; the rest was for showers. We took "Navy showers" on the ship, which meant we wet ourselves down for fifteen seconds and then lathered up. After we soaped up, we turned the water back on for a minute or so to rinse off. The only thing we had to worry about in the shower, we joked, was dropping your soap. A sailor who was filthy or greasy, or hot as hell, turned the large red handle on the side of the shower stall and enjoyed the clean, ice-cold Pacific Ocean. There was no privacy in the ship's showers; no curtains hung in the shower stalls. Anyone wasting fresh water at sea was immediately confronted by another sailor or two. Self-policing in a few institutions, like the military service, actually worked. We knew and obeyed the rules, most of the time. Most sailors celebrated coming home after a cruise by taking a long, hot shower just because they could. That shower and going out for a favorite meal with friends meant we were home.

Sailors called bathrooms "heads." All sinks were shiny metal, but everything else in a head was painted battleship gray. Sinks were often covered with beard stubble left by sailors who dry-shaved and didn't have the common courtesy, or proper upbringing, to clean off their mess. The toilets didn't have doors. I became so used to doing my business in front of twenty different sailors that I could defecate into a bucket while reading the morning paper as I sat in the window display of Macy's Department store in the Big Apple during the Christmas holidays.

The heads were usually fairly clean; we took care in picking up after ourselves. But occasionally there were leaks in the ship's plumbing, and the floor had a layer of water on it. When the ship rolled in heavy weather, the water on the floor in the head slid from side to side, forming a puddle when it hit the wall. At those times, a sailor using an end toilet sat in an inch of water until the ship rolled back the other way, and then his commode sat on a dry floor once again. Or rather a damp floor, but the bottoms of his pant legs would be wet had he been too slow to raise them above the rising water.

What pieces of work we sailors were; on liberty, we drank much of the day and night, went to bed for ten hours, and as soon as we got up, we were right back at it. One morning after a full night of carousing around Waikiki, Pete and I painfully crawled to our feet, went to roll call, learned we had the day off, and headed back to our racks. In the early afternoon when we finally rolled out of our sacks, we went to town to look around and get a beer. In Honolulu we found plenty of beer, but there were no women on the street; the few working women were already engaged by the time we got there. That was a blessing. A misconception about navy life was that sailors always went ashore to interact with women. The truth was sailors could no more escape being around loose women in navy towns than a preacher's wife could escape being around mosquitoes at a summer picnic in Savannah, Georgia. Almost every sailor went ashore only to sightsee or have a cold beer. In a navy town, loose women hanging around sailors were like fleas on a dog; you wouldn't find one without the other.

Honolulu was a tremendous navy town, but it wasn't wild. It was like moving Charlotte right up to the ocean at the beach in Hilton Head. The only women we encountered downtown were beautiful Asian girls who were available but not for common sailors like us. They were way beyond the budget of any sailors we knew and were not accessible for public consumption; they were prime Kobe beef. All we wanted to do was to talk to someone who was not a sailor; someone more like an attractive female. We saw many ordinary and extraordinary, plain and fancy, fat and slim Hawaiian girls and women, and they were all sitting on bus benches quietly waiting for their rides or buses to come take them home after their long day's work. We sat and talked with them at times, but we considered them as unattainable as nuns in a convent, yet as desirable as Krispy-Kreme donuts.

The loveliest woman we met during our entire cruise was in a nightclub in Hong Kong that she used as her "home office" for her ultra-exclusive small-business escort service, and she became quite taken with Pete. A four foot framed poster of William Holden and Nancy Kwan from the movie *The World of Suzie Wong* hung on the wall near the entrance. Much of it was filmed in that club eight years earlier. The beautiful Asian call girl sat and chatted in perfect English with us for a while until the manager came and beckoned her to the phone. Some rich Japanese or American tycoon or politician requested her valuable company; it was time for her to go to work, but she was nice enough to come back and say goodbye. It was just a simple matter of supply and demand, something we would see the rest of our Pacific deployment. Another beauty I especially recall was a twenty-something Japanese woman with the most beautiful green eyes I've ever seen. I stood to give her my seat on a city bus; something Japanese men never do. I still remember the look she gave me as she wordlessly thanked me.

At sea, a ship's crew prayed for the peace and quiet we shared on those Honolulu benches, but sometimes we received the opposite. On one occasion in the middle of flight ops on Yankee Station three months later, Pete and I were caught up in the utter madness of the readiness condition called General Quarters. The action mobilizes the entire ship, endeavoring it to reach its maximum ability to defend or defeat some imminent danger, either from an enemy or from within, by directing all personnel to their battle stations. We were below decks telling the computer experts responsible for all non-factory repairs to the A6A central computer unit, the brains of our system, what the bombardier had noticed about this particular computer on his last flight. We had pulled this computer drum, its brains, out of his plane and replaced it because it intermittently stalled and flickered, leaving the plane without a viable weapons program. Our job was to explain the negative functioning of the inconsistent computer operations to the men who had to fix it. Pete ran over to me when the ship's gongs began booming.

"Jesus, what's that? What's happening?" Pete could hardly talk.

One petty officer in the AQ repair shop said, "It's General Quarters; it's bad news. You need to get to your shop. Now, before they seal you off."

We tried but couldn't get up on the hangar deck before all the doors and hatches were locked and sealed. We had rushed in a wild panic up and down a maze of open ladders, stairs and passageways, trying to avoid the screaming flying squads, and had unbelievably found ourselves in the middle of the ship. We must have been the dumbest men on the ship; nobody else was running around like chickens with their heads cut off. Had they already found their battle stations? Did we even have one? Was that information in those 'emergency' flyers we saw pinned to the bulletin board in our shop; the ones we never read?

We decided to stay wherever we were; we couldn't go any farther. We soon found we were stuck in a corridor near the infirmary, or sick bay, when two men, running quickly down the narrow hall, squeezed tightly side by side, approached us. Cradled in their arms was a young black sailor who had been on the hangar deck at the mouth of a jet engine being repaired when it cruelly flamed out. The searing flames spurting from the engine burned the young man severely on all exposed places of his body. His long-sleeved color-coded sweatshirt, bell-bottomed pants, work shoes and hat had saved most of him from the horrible fire, but the burns to his face and hands were horrendous. His wounds were very obvious because his skin was a deep black color where it was not burned and was white as snow where it had been. The edges where the white skin ended had a hard, brown, scabby crust all around them.

I looked into the face of one of the men carrying him. "Is he dead?"

"Not yet, but it's like he's been in an oven," the boy said. "Can't talk; gotta go."

As the severely injured sailor was carried into sick bay, Pete and I got a close view of his maimed body. Because the passageways were so narrow, he was pushed against us as he was rushed by. Sailors had to turn sideways to pass without contact in many areas below decks. I noticed his partially white face was splotched with little rosettes of pink. He looked unconscious, and it appeared to my untrained eye he wouldn't survive his injuries. Thirty feet down the passageway, the three sailors turned into the infirmary, gone from our sight. In just a day or two later, the sailor was forgotten by us, gone from our minds. But not because he was black, no, not at all. Pete and I were just simple, self-centered young men working hard to do our jobs, and we forgot him like we forgot everyone else. We never asked any authorities, and it was never reported by any, if he lived or died.

"Anybody heard anything about that black guy who was burned in the flameout?" I asked the guys in my shop the next day.

"What black guy in what flameout? Oh, during GQ?" Joe Houseman, a new guy in the squadron, replied.

"Good God, man, just forget it. You ever been to sick bay?" I asked.

"Only once, they made me get another shot for the cruise," he said. "You?"

"Yeah, I've been one time. We were washing planes in Hawaii. I was up on the back of mine and decided to take my shoes off. Our plane was covered in cleaning fluid, and we were standing on top. I moved off the centerline to swab near the side and began sliding when some asshole started shooting water up on the top. Chuck Ross reached out and stopped me with one hand. I stood there perfectly still for about a second and said thanks, I'm okay, let go. He did, and I immediately slid off the plane onto the concrete tarmac, or whatever it's called, feet first. I tried to soften my landing and rolled backwards like a paratrooper when I hit, but without my shoes it still hurt like hell. How high up is the top of a plane? 15 feet? Well, I got my shoes on and hobbled back to where a cattle car would take me to the ship. I told the guys hanging out in the shop that I was going to lie in my rack until the pain got better. The next day, we were at sea and I stumbled down to sick bay. There's not one damn place to sit while you wait, so I stood there in line in the passageway. I guess all the guys were getting their VD shots. After a while I gave up and went back to my rack. The Chief sent me some aspirin, and the next day, still hurting like hell, I went back to work."

"You never got anybody to look at them?"

"Hell no. You know what they say, steel doesn't break; it bends."

When a sailor died at sea there were options as to what to do with his body. Most times it was stowed on board until a plane left the ship with departing personnel or outgoing mail. That sounds so callous, but if anyone comes up with a better, more efficient way to get a dead sailor off a ship, the Navy would want to hear it. Departing mail shipments happened almost daily, so any deceased sailors or Marines were quickly returned to their loved ones or to a military cemetery. Sometimes the sailor's family, after it has been notified of his death, decided to have their son buried at sea. No general notification or announcement to anyone of the ship's crew, or to any personnel of the air wing, was made about the upcoming service. The deceased sailor's close shipmates knew of the service because they worked with the man and wanted to pay their final respects to him. Other personnel watching the service just happened to walk by and notice it for what it was.

Two days after the death of our forgotten brother-in-arms, Pete and I were walking to the chow hall at the bow of the ship. As we passed an elevator on the hangar deck, we heard a soft murmur coming from a crowd milling around in the sunlight. Pete asked me, "What's this, a church service? At one o'clock? It's not even Sunday, is it? Look at all the people."

"I don't think so. I think it's some kind of funeral; that's what it looks like. Man, this is something! You think that black boy burned in the flameout died? Wait…hold on a second, Pete, I want to see this," I said, grabbing him by the elbow.

We stopped and looked around, and then I spied the ship's band assembling near the far side of the elevator. I thought back to the crowded passageway the day the young man was burned and said, "That poor boy was pitiful, and I felt ashamed or something. I was kinda embarrassed when I saw how burned he was. I didn't want him to touch me, but I couldn't help it; I couldn't get out of his way, it was too crowded. I felt sick seeing him like that."

"You didn't want him to touch you? Is that what you said? Was he dirty or something? Was his blood or his pus going to get on your jacket? I can't believe you said that."

"No, that's not what I meant, goddammit. Don't start with me, Pete…I didn't want to get any grease or dust in his wounds from my jacket. From my jacket. I didn't want to infect him. You know that."

"Yeah…I guess…let's get closer."

The brilliant sunlight coming into the ship from the opening above the elevator made everything glow with a dazzling shine. The carrier's ensemble of twelve musicians, our little drum and bugle corps, waited on one side. An eight foot long wooden box painted battleship gray was balanced on an old metal gurney in the middle of the elevator, surrounded by six sailors in dress whites wearing snow-white gloves and standing at parade rest. The ship's chaplain rose from his folding chair, and at a nod from his uncovered head, the band began playing

"Nearer, my God, to Thee," and an old immaculately dressed Master Chief slowly stood up and began singing in a rich baritone. A couple of other hymns I remembered from my younger days in church were then rendered flawlessly by the group; "The Old Rugged Cross" and "How Great Thou Art." Good old Methodist funeral hymns.

We edged ever closer to the middle of the bystanders and watched as the chaplain finished a prayer. I wanted to see that marvelous ceremony. I wanted to touch that painted box which surely held the dead sailor. I wanted to tell him how sorry I was that he died so young. I wanted to bid him goodbye. The mini-band started playing "Anchors Away," and eight Marines snapped to attention. One held the ship's ceremonial flag with all its beautifully embellished streamers gently fluttering upright, lifted by the light ocean breeze. The other seven raised their rifles and fired a shot into the distant horizon of the Pacific. The roar of the simultaneous blasts startled everyone on the hangar deck. The Marines dropped their rifles to parade rest and after a second, raised them again and fired a second volley. Everybody jumped again. The hangar deck felt small and crowded with all the jets parked nearby. The discharge of the rifles echoing off the metal bodies of the planes gave me chill bumps. One more time the rifles were lowered and then raised, and the last of the seven rounds was fired. The three volleys of seven shots each produced a marvelously performed twenty-one gun salute on an aircraft carrier at sea in the middle of a war.

As the last of the rifles' retorts faded away, the lone Marine yelled, "Attention," and we all froze where we stood. The six sailors in their dress white uniforms stepped along both sides of the wooden box, lifting it two inches above the hard metal gurney. Holding it carefully at waist level, they carried it twenty feet to the edge of the deck. The sailors at the back slowly raised its closed rear end, tilting the box's hinged front end toward the ocean, opening the door and releasing its contents. The plain gray box had held our dearly departed sailor, someone's beloved son or brother, until it was time for his body to leave this world. The ceremony had been performed for over four hundred years, countless times, but with never more precision or care. The young man was clad in pristine white swaddling clothes and looked like a mummy in a shroud. He shot out of the box and flew downward through the salty air until his body slammed feetfirst into the cold Pacific and quickly sank to Davy Jones' locker. Following ancient naval tradition, the last stitch sealing the shroud had gone through his nose to ensure he was dead, and the shroud had been weighted down to prevent his lifeless body from floating to the ocean's surface.

That young man, that fine African American boy, had one of the best funerals money could buy. He gave up his body and soul in the service of his country to receive that richly deserved and great honor. Only God knows whether it was worth it. The gawkers and ship's personnel disbursed, leaving about a dozen guys I considered his buddies huddled together, saying a final prayer of their own. One was holding a large picture of the boy, immaculate in his dress blues, blown up just for that service, and he was indeed the boy we had forgotten.

We walked over, and taking a long, close look at the picture, I asked the sailor, "How did he die?"

"He was burned in the flameout a couple of days ago."

"Yeah, I know that, but what I meant was what finally killed him; the burns, infections, maybe his heart?"

He said, "No, his lungs. They were almost totally incinerated when he took his first breath in the flameout. The docs say he wouldn't have made it if he had been in the best hospital in the world. You guys know him?"

"Yeah…but not too well. We only met him once."

Moving on to the chow hall, we passed the most beautiful jet on the ship, an RA5C Vigilante, and Pete said, "Man, I'd love to fly one of those mothers."

"In your dreams," I said. "You know, those things don't even carry weapons; they just take pictures."

"You're kidding. No weapons? What a waste."

"Yeah…Man, I hope they got those real potato French fries today. I hate that powdered crap they squeeze through that baker's icing cone. They just fall apart when you pick one up."

"Are those things even potatoes?"

"Who knows? When we finish eating, I've got to run down to the ship's store and get some throat lozenges; my throat's killing me."

"That's 'cause you've been licking the Chief's butt too much."

"Ha; look who's talking. You're the biggest brown nose on the whole screwing ship."

"Screwing? Did I ever tell you about screwing your girlfriend?"

"You mean your mother?"

8. AIRCRAFT CARRIERS

Honolulu, Waikiki, mahi-mahi and the surf were terrific, but the time had come to engage the enemy. A few more days in Hawaii, topping off our oil supplies for the ship's boilers and our aviation fuel supplies for our planes and cramming food, ordnance and material supplies anywhere and everywhere, and we would be ready to ship out. I was huddled in my rack trying to block the pain from my swollen ankles and ignored everything else, everything except the head once a day. Some of my friends brought me some non-messy snacks and water a couple of times. That's the only time I was sick or injured or tried to go to sick bay the entire four years of my Navy days. We were to sail for Japan that August of '69, beginning *Connie*'s fifth combat deployment. Carrier Air Wing 14 consisted of fighters and bombers from squadrons VF-142 and VF-143 flying F4Js, squadrons VA-27 and VA-97 flying A7As, and our squadron with A6As. Her non-combatant aircraft were an RA5C supersonic reconnaissance jet of squadron RVAH-7, an E2A early warning aircraft of squadron VAW-113, and a KA-3B tanker from squadron VOQ-133. Helicopters were attached from squadrons HO-1 DET 5 and HO-7 DET 110, both with SH-3A choppers. We were scheduled to begin twenty days of initial air strikes in South Vietnam and Laos as soon as we arrived on Yankee Station. It would be just a matter of days before every soul on the *Constellation* would be operating in a combat zone.

When the *Connie* got out into the open water of the Pacific, the non-stop, full-speed-ahead sailing time from North Island to Pearl Harbor would be almost exactly three days. Three days. It could be done in an emergency, but we were in no hurry; we would even conduct flight operations. We had our schedule, and being shipboard and in the military, we stuck to it. The two thousand six hundred and fifteen mile trip sailed at thirty-six miles per hour would take seventy-three hours, or three days, non-stop. And on the open sea, she usually sailed non-stop, night and day, between ports. The only variations made to course and speed were to facilitate the launch and recovery of our aircraft during flight operations. Practicing touch-and-go maneuvers, which were landing and taking off without stopping, sharpened the skills of our new aviators during the voyage to Hawaii. Our aircrewmen needed to get back into the feel of shipboard life, living in an environment completely enveloped by iron and steel, after almost eight months on shore.

An aircraft carrier engaged in naval aviation operations at sea is a living, breathing monster, a gathering of shipboard sounds of banging, and clattering, and screeching. The whining of the powerful jet engines as they spin to life drowns out all other sound. Communications between crewmen during a launch of aircraft were made by hand signals due to the horrendous din on the flight deck. At night the signals were made by using yellow lighted wands.

Pete and I were computer and radar technicians, not ship's ground crew, so we weren't supposed to be on the flight deck during flight ops. But we were able to catch a deck-level view of a launch by hunkering down in a catwalk in front of the tower, which was highly against regulations. The policy was made to keep pseudo sightseers, unnecessary and untrained sailors like us, out of the way of the serious business underway. The catwalks were forbidden to us during ops; a no-man's land where a fire or an explosion could kill anybody. Standing in the middle of gurneys filled with bombs waiting to be raised up unto the flight deck, we had to act like we were working. To quicken the ordnance loading time, bombs were also stashed on the flight deck all around the seaward side of the tower.

The highlight of a launch cycle of twenty or more jets in one wave was the yellow and orange flaming cones screaming out of the back of a jet's engine when the afterburner is used. A jet needs an additional burst of thrust and speed when taking off from a carrier's flight deck. The afterburner is the part of a jet engine that injects additional aviation fuel into the gas stream behind, or after, the turbine. Afterburners increase the jet's airspeed, so it can achieve and maintain its initial flight. The part of the deck used for take-offs is only the front half of the ship for the forward cats and the entire angled deck veering off to the left, or port, side of the ship for the side cats. The ship's catapult system is hidden beneath the flight deck and pulls the jet off the ship. When a heavily-laden jet leaves the carrier, it slopes down below the line of vision from the middle of the flight deck looking forward and is lost from sight. In a matter of seconds, the jet can be seen again as it rises from below the front of the ship and roars away. Occasionally, it disappears for good and doesn't roar away.

To increase the wind across the ship's bow, the carrier is turned into the wind at launch time and increases its speed to maximum. If an aircraft does not have enough airspeed to enable flight, it crashes into the sea, and a command is given that a plane is down and in the water. The ship's navigator in the tower knows to immediately turn the ship to avoid running over the downed aircraft and its pilot or flight crew. Helicopters and a destroyer follow close behind and to the side of the carrier whenever it's engaged in flight operations to quickly rescue any aviators from a downed aircraft. Aviators turn their disabled jets to the port side, eject from their cockpits and are hopefully picked up at sea. The downed aircraft are so heavily loaded with fuel and bombs and other weapons of war that they sink immediately.

Everything on a carrier during flight ops is done for one objective, to increase the wind speed across the jet's wings to provide the lift needed to enable the jet to fly. Not any jet, but a fighter or attack aircraft weighing tens of thousands of pounds and carrying thousands of pounds of fuel and thousands of pounds of assorted rockets and bombs. Our VA-85 A6s weighed twenty-six thousand pounds empty and sixty thousand pounds fully loaded with fuel and bombs. The only naval aircraft that weighed more than ours was the RA5C Vigilante at sixty-six thousand pounds. And she carried no weapons; she was a reconnaissance plane loaded with cameras. Imagine the force and power necessary to hurl a sixty thousand pound death-machine off the flight deck, accelerating from zero to one hundred and forty miles per hour in just three seconds. From an absolute standstill of zero to a takeoff speed of one hundred and forty miles an hour in three seconds! Sixty thousand pounds of death and destruction. And that was fifty years ago; modern nuclear carriers launch their jets at one hundred and sixty-five miles an hour today. In three seconds.

There are four different catapults on each carrier's flight deck. Two are on the forward deck heading directly toward the front, or the bow, of the ship and the other two are on the angled deck. The angled deck launches its jets off on a line between the ten o'clock and eleven o'clock positions of the ship, assuming the ship is heading toward twelve o'clock. Under that setting, two jets can be launched almost simultaneously, one directly ahead of the ship and one slightly to its left. Having two catapults on each of the two launch pads allows the catapult crews to prepare two jets for launch in both directions at the same time, making it a very efficient system.

Our catapults used pressurized steam accumulated in cylinders from our oil-heated ship's boilers; on modern carriers the steam comes from the heat of their nuclear reactors. The steam is then released on command from the four separate cylinders below the flight deck, each as long as a football field. The steam propels the aircraft the length of the catapult using a cable attached to the nose wheel of the jet, and a pin shears off and falls away at the end releasing the jet. The *Connie*'s catapult system's launch limit was a seventy-eight thousand pound load at one hundred and forty miles an hour in three seconds.

After their missions, the aviators fly their jets back to the ship for recovery. The jets circle the ship and line up at its rear end, the stern. A landing signals officer guides the aviator in, advising him of adjustments needed for height and angle corrections. Jets land, or are recovered, on a carrier using the angled deck and only while the jets are under full power. The use of the angled deck ensures the plane will have a clear deck and can maintain sufficient speed to touch and go, if needed. That maximum speed is important so that a jet that misses the ship's arresting cable can easily lift off the deck and continue flying again. Within seconds of the planes being captured by their tailhooks on one of the ship's four arresting cables, they are released and gathered at the bow of the ship.

Our squadron's sleeping quarters were just below the flight deck, right where the arresting cables were stretched across the deck. Every time a jet landed, its terribly heavy tailhook slammed onto the thick steel of the flight deck, jarring us in our sacks. As the cables played out, gradually stopping the jet, the catapults whined like a Saragosa spinning reel as a Marlin makes its run, or a NASCAR engine as it tops two hundred mph at the black-and-white checkered flag. As a sailor lies there waiting for the next jet to land, the roar of its jet engine can be heard as it approaches the rear of the ship, even before it has actually reached the flight deck. The pilot increases his speed to full power in case he misses the cable. When the tailhook hits the deck, the noise is like an automobile dropped onto the deck from a four-story building. Then, as the plane stops, backs up and releases the cable, the catapult rewinds and gets ready for the next trap, or capture, of a plane.

It whines again as it coils itself back into its tube, or cylinder, and the pilot races his jet engine to overcome inertia and move out of the flight path. Recoveries only last forty minutes to an hour if there are no problems, and it's a lie that sailors sleep so well in their comfortable, thin fart sacks that all that unbearable noise is hardly noticeable.

If the aircraft has been on a long mission, or if the pilot is tired or distracted, or if the ship is heaving too much in rough seas, the aviator may miss the cable more than once. If fuel is low, and the weather is bad, and no base is near, the situation becomes acute; with no other recourse, there would be no other place to land that jet except the deep blue sea. Fortunately, carriers stretch an emergency barrier, a gigantic net, across the flight deck to grab any jet that cannot make a normal landing due to damage or lack of fuel.

During a recovery, if an arresting cable snaps, causing a jet to veer off course and doesn't stop it, the other jets lined along the side of the flight deck are easy targets to be rammed. If that happens, the rushing, careening twenty-six thousand pound jet is transformed into a projectile of unspent fuel and any unjettisoned ammo, destroying anything in its path. Anyone in, or near, any of those jets along the edge of the flight deck would be in extreme danger. The most dangerous job in the Navy, and possibly the world, was manning the arresting cables on a carrier in the '60s, and it may still be. If a frayed cable snapped from the jet's force funneled down and concentrated into its tailhook during a recovery, it would sling like a horsewhip across the flight deck and sever the legs from anyone in its path.

Ignoring regulations, Pete and I once or twice purposely forgot to go below when our recovery ops began, leaving us inside a jet on the edge of the flight deck as other jets landed. That sorry-ass forgetting excuse wouldn't have saved us from a court-martial had we been caught. Only simple, last-minute, but essential computer, navigational or communication repairs could be done on a plane waiting on the side of the flight deck during flight ops. The cockpit seats in our parked A6 provided us with a wonderful view of all the action. Besides the breaking of arresting cables, the flameouts of jet engines, the propellers of our prop planes, being sucked into an engine, and being blown overboard, the thing we most worried about was not accidently discharging the pilot or bombardier/navigator's seats' ejection system, thereby being blown a hundred feet into the sky or the top of the hangar deck. The explosion wouldn't kill you, but blasting through the canopy and landing in the sea might.

I asked Pete one day, "What's your favorite thing to see or do on the ship?"

"Man, it's got to be a night launch...in person. The only thing better than seeing a day-time launch is seeing a launch at night. Gives me goose bumps. All those crazy flames and sounds. Nothing but pure excitement."

There were no bright lights outside an aircraft carrier at night; everything was illuminated with dimly-lit red lights to protect our night-vision. A rookie cutting on a flashlight would soon wish he were back home and not at sea. The primary light sources at night, excluding jet exhaust, came from the green and red lights on the jets lined up for launch and the yellow wands of the launch personnel. Under their own power the jets eased themselves above the cat pad, and once secured to the catapult cables, they were ready for take-off.

A large metal wall known as a Jet Blast Deflector, or JBD, was hydraulically raised as each jet sat on the cat maximizing its power using the jet's afterburner. Additional fuel poured into the afterburner behind the turbine, but still inside the engine cowling, created the beautiful yellow-orange flame spewing like brimstone from its wide-open mixer fans. The giant glowing cone of fire shot from the rear of a Phantom twenty feet until it hit the two-foot thick and ten-foot tall pneumatic wall behind it. The flame then climbed up and over the top of the deflector for about five feet, still as a cone. It then disintegrated into a vapor of boiling hot jet engine exhaust. The afterburner's flame and cone were visible in the daylight, but it was nothing like its dynamite performance at night. The forty-foot wide JBD was used primarily to keep rockets and missiles, armed and waiting on the jets to be launched, from cooking off due to the afterburner's heat. It also protected waiting jets' fuselages and prevented any aircrewman standing too close behind the jets from being fried, a mistake only made once.

"You ever see anybody cooked by a jet's exhaust, Chief?" Pete asked one day.

"No. Whatta think those damn JBDs are for, jackass?"

Other dangerous aspects of flight ops were the possibility of being blown overboard, or at the opposite extreme, from being sucked into an engine. The flight deck of a carrier was gigantic, but planes were all over it. When the aircrew was preparing for a launch, jets were queued at the back half of the ship to be led to the catapult. As they launched, another batch came up on the elevators from the hangar deck. Those were spotted so the plane handlers could park the jets where they could reach the cats under their own power.

The movement of twenty jets back and forth and from side to side with their turning and spinning to get into position for launch made it easy to get in the down blast of a jet engine. Flight deck personnel were constantly on alert to the moving jets around them. Catwalks all around the edge of the flight deck were large enough for a man to walk in and wide enough to catch someone falling off the flight deck. But they did not always stop someone from being blown off the deck like a rag-doll. No safety nets stretched below the flight deck to save anyone from the cold Pacific Ocean or the Gulf of Tonkin. In combat situations a carrier itself cannot respond to a man overboard; there are more important things to worry about than one sailor's life. Hopefully, the ship's helicopters or the destroyers sailing around and behind the carrier would rescue the sailor; that is, if someone had noticed in the confusion of the moment that he was actually blown overboard. Twenty-one sailors would die before the *Constellation* returned from this Western Pacific cruise; none of them blown overboard, but many had during flight ops and major fires in the past.

As the jets spun around and moved into position, their jet engines were fully operational, and their fans were loudly whining. The jets screamed every time more fuel was applied to break the hold of inertia and move them where they needed to be. The noise was horrendous, but it was marvelous; it was stupendous. It was like being inside a jet engine. The danger was present even before the jet was pinned to the catapult and the afterburner switch was flipped on. The jets sucked air into the engines where it was compressed and then ignited, and the powerful thrust of the exploding exhaust gases inside the engine pushed it forward. The engine was a part of the aircraft, so the entire aircraft was pushed forward. All the engines of the jets on the flight deck were started five to ten minutes before they were launched. Any man standing up just in front of the engine's cowling was easily sucked off his feet and pulled inside the engine to be eviscerated by the turbine's spinning blades. I never read that anywhere, in school or on the ship, and was only told it once. I was on the flight deck for the first time, walking around, looking for an exit, and a friend took me aside and warned me. That was something a sailor should never forget, but some did. With twelve to twenty jets roaring at one time, it was hard for a sailor to tell if the one just in front and just behind him was started. To keep from dying, a sailor had to assume every jet engine and every propeller on a prop plane was turning until he found out otherwise. Again, he would only make that mistake once. An engine's exhaust debris at that point was one of blood and tiny bits of flesh and cloth blowing out its back. Structural damage to a jet from a blown engine could cause an even greater secondary explosion or fire. Aircrewmen grabbed their firefighting equipment, always on hand, and rushed to the plane to extinguish any leaking jet fuel before a major fire and possibly doomsday. The danger went beyond the burning plane; it encompassed the whole ship. The loss of one plane meant nothing except the loss of twenty-five million dollars.

"Twenty-five million dollars? Just one plane? My God, how big is our budget? What's this war costing us? Think of how many hungry kids could be fed with that kind of money," Pete said when he first heard that figure.

"Yeah, but think of what those kids would be eating without our big military budget," I said.

"Whatta mean?"

"They'd all be eating sweet little German strudels and 'sprechen sie Deutsch' if the Nazis had won. That was one helluva army they had, and they'd be our rulers right now…Or maybe we'd be eating that Russian beet soup, because those bastards would have done the same thing. Their armies were even stronger and did more to beat the Nazis than we did," I answered.

"You don't really believe that crap," Pete said.

"I certainly do."

All jets were filled with a very volatile jet fuel which could flood the deck if a wing or fuselage tank was ruptured or an engine was blown. The loose fuel could catch on fire, burning other jets and all fueling equipment on deck, causing an inferno which might not be extinguishable. In addition, during combat operations, each fighter/bomber was armed with missiles and bombs which could be detonated by explosions or fire. Once ignited, one of those missiles could slam across the flight deck to hit another jet, activating its missiles. A lot of ordnance would be on deck and in the catwalks and vulnerable to ignition; a hot fire could cook them off. Fuel could run along walkways and through maintenance holes and hinges on the flight deck and could slip down into the hangar deck and the sleeping quarters below. One burning plane could become a towering inferno creating a vicious deathtrap similar to the *Forrestal* in '67 when 134 sailors were killed.

All sailors who had duties that were performed primarily on the flight deck were trained and practiced, again and again, on fire control and ship safety. All other sailors were instructed on those same skills in firefighting school, and although the picky details were soon forgotten, the essential actions and movements involved in fighting a jet fuel fire never left you. Ordinary sailors on deck during a fire were there only to assist the firefighters, unless the situation turned deadly and they were forced by injury or death to take command.

The most important lesson a sailor on a military ship learned was to get out of the way of the "flying squad" when General Quarters was set, or any other serious incidence occurred. The flying squad members were hand-picked experts in dealing with every possible emergency aboard a ship. When the alarm was sounded, and the ship's gong started booming, the flying squad was called to instant duty over the ship's loudspeakers.

"Away the flying squad. Away the flying squad." Bong, bong, bong.

Those brave men immediately dropped anything they were doing and raced to their assigned repair locker which held their firefighting gear. They took off and flew down the ship's corridors and leapt down the metal stairs connecting the different decks of the ship. When they reached the top of a stairway, which resembled a narrow non-moving escalator at the mall, they lightly slid their hands along the rails, barely touching them, and without using any of the steps, hit the bottom twelve feet below and took off. They did not stop for an instant. The only thing the other sailors were ordered to do at that point was to get out of their way and let them do their job.

"Gang way. Out the way. Clear that passageway. Move that crap. Now, sailor!"

The command "General Quarters. General Quarters. This is not a drill. This is not a drill," accompanied by the constant bonging of the ship's sound system scared the hell out of everyone on board. It was then repeated with an announcement of what the emergency was. "Fire on the hangar deck. Fire on the hangar deck." General Quarters was the most crucial command given on a warship; more crucial than "Abandon Ship. Abandon Ship." General Quarters required sailors and Marines to go to their individually designated battle stations without interfering with the mad dash of the flying squads and firefighters. All the ship's airtight doors and lockers were sealed, and everyone awaited further instructions. During General Quarters, sailors were trying to save a ship in deadly peril. During Abandon Ship, everyone was just trying to leave a sinking ship.

9. JAPANESE HOT TUBS

Our first orders putting us into the Western Pacific combat theatre sent us from the lovely islands of Hawaii to the U. S. Naval Base at Sasebo, Japan. From there we would shortly take our place on Yankee Station in the Gulf of Tonkin. Before our final farewell from Hawaii, our supply officers took on as much fresh food as they could commandeer, much to our great pleasure. Getting fresh milk, the finest thing we were ever served on a ship, was the highlight of any port call. Within three or four days at sea though, it was gone, and we switched back to lukewarm powdered milk and yellow or red Kool-Aid straight from a hot mixing tank. There often was no ice to cool it in the enlisted mess, or possibly the Chiefs' or Officers' mess.

One special chow hall served hotdogs, hamburgers, chili and fries almost twenty-four hours every day of flight ops, but not in port. It wasn't healthy food, but no one cared. And it wasn't always that good, but no one cared too much about that, either. Besides, there was always a bottle of Heinz hot cocktail sauce and Tabasco on every table to smother the food. I seldom ate an egg, hamburger, hotdog, or casserole while on an aircraft carrier that wasn't jazzed up with a load of cocktail or hot sauce. Most everybody I knew did the same. And the beauty of eating in any military chow hall was the cooks gave you more the second time than they did the first. That was the military, or at least the

Navy, way. Apparently, the cooks thought if you were stupid enough to want some more, you could have all you wanted.

I was amazed watching the cooks fry eggs in the morning. In port, we got fresh eggs, and we loved them. One or two of the cooks started cracking eggs and dropping them on the two foot by ten foot griddle. Two or three rows of ten to fifteen eggs each were usually cooked at a time. By the time all twenty or thirty eggs had been dropped on the grill, it was time for the cooks to start flipping them over. When the last one had been flipped, the first ones were ready to serve. As each man walked past, the cook would throw one or two fried eggs on his plate. I cringed whenever one of the eggs sitting on the grill had a yolk as black as the Ace of Spades. It stunk like rotten sulphur, but the cook just cut it from the others with his spatula and slid it into the grease trap. If you wanted scrambled eggs, you just yelled "Scrambled" as you came to the edge of the line, and the cook would gather two or three together and break their yolks and chop them up. When we didn't have fresh eggs, we were served powdered eggs. The only way to stomach those was with an almost equal part of that hot chili sauce.

However humble our food was, there were some men who thanked the good Lord for our chow. Occasionally, we saw a soldier or a Marine sitting alone at a table in the chow hall, eating from a metal tray loaded with the day's delight. The men stuck out like sore thumbs because they would be dressed in their olive drab uniforms and would be leaning over their trays like they were trying to prevent someone from stealing their food. Those men had that distant look in their eyes, and we knew not to bother them; they looked like they were dangerous. We never said a word to them. They always had their free arm wrapped around their trays defending them from attack. Pete and I thought they had come on board for a little rest and relaxation before they cracked, but some medical procedure or treatment was the most probable reason. Many soldiers and Marines serving in-country in Nam would think it was heaven to live on a large ship for four or five days, eating three hot meals a day and sleeping in a cool bed in a fairly safe place.

We had a couple of crazy cooks in the chow hall where we usually ate. They were just fun-loving, young, bored guys who were often looking for a laugh when they weren't swabbing decks, peeling potatoes or scrubbing giant pots and pans. They served us in their delightful and sexy working attire. Whenever we had fresh fruit for breakfast, they stuffed their soiled undershirts with a couple of large grapefruit or oranges to make it look like they had enormous breasts. Those greasy undershirts were the only thing we ever saw them wear. Then, they adorned their heads with the loose strands of a mop and an enlisted man's cover, or hat. Sometimes they smiled at you, revealing their snow-white teeth separated by four or five they had blackened. Those boys helped lighten our load. Of course, we only had fresh fruit while we were in port, so those boys may have still been drunk from the previous night's liberty.

As we approached Japan, our Navy brass was worried about our reception in Sasebo. The U.S.S. *Enterprise*, the first American nuclear powered aircraft carrier, docked there a year and a half earlier, and many demonstrations and harsh clashes occurred between Sasebo security and Japanese activist groups. No one in Japan, other than the hordes of merchants looking for the mega dollars spent by an aircraft carrier's crew in port, wanted a nuclear carrier anywhere near their country. Tens of thousands of Japanese students had marched, and many were jailed and hurt. Nearly twenty-four years had passed since the bombing of Hiroshima and Nagasaki during World War II, and the Japanese still felt the bombing of those cities was unnecessary and excessive. They felt unjustly victimized by America, but many reasons justified the detonation of one or both of the only two atomic bombs we possessed.

While waiting to recover our planes one evening, we were in the chow hall drinking coffee, discussing our upcoming port call in Japan. Pete asked no one in particular, "What's the news on the Japs and our docking this year? They going to protest and burn flags and all that crap?"

A sailor who looked to be a teenager was sitting by himself at the end of our table and made the mistake of answering Pete, "I heard five or six thousand are going to show up, and I don't blame 'em; why did Roosevelt have to drop the atomic bomb on all those civilians?"

That really griped Pete's butt, and he said, "Look, buddy, I don't want to discuss it with anyone who doesn't know anything about it. Where'd you go to college? You some hippie?" Pete was being a smart-ass; he knew the boy was too young to have been in college.

The boy said, "Hey, I didn't go to any college; I was a high school dropout until I was drafted, but I still know my American history."

Pete rolled his eyes and smiled; he was going to have a little fun. "Okay, but getting drafted doesn't mean you're not a dropout, Einstein. They normally have nothing to do with each other, except in your case, it may actually have been the reason you couldn't avoid the draft."

The boy looked over at Pete. "How about you? Why couldn't you avoid it?" the boy smirked; his hackles rising.

"I wasn't drafted, I volunteered, but that's another story. My mother had terminal lung cancer, and I couldn't just get up and run off to Canada with all the wimps trying to escape the draft; could I?...Besides, she needed me...I was the only person who'd go to the store and buy her some cigarettes," Pete chuckled and gave me a high-five. "Not really, man, I'm just kidding about that; lung cancer is a serious business. But here's what really happened. Franklin Roosevelt died in April of '45, and Germany surrendered that May, and Japan surrendered in September. But to get Japan to surrender, our guys had to decide whether to invade Japan and lose a million of our men or to drop the bombs, two bombs, killing three hundred thousand Japs who started it.

Okay? That's why we dropped the bombs, but Roosevelt was dead, so Truman did it; God bless him."

American Army and Navy planning departments dedicated months of calculations to determine casualty and death estimates for an Allied invasion of Japan. Operation Olympic would hit the beach at the southernmost Japanese island and be followed by another beach invasion near Tokyo. Those were our only possible invasion sites; the Japanese knew it and were prepared to fight there to the death. The invasion forces would be six times the strength of the forces that invaded France on D-Day. The estimate of Allied deaths was seven hundred thousand with total Allied casualties of four million. The total Japanese casualties of an invasion was estimated at ten million. Weighing those estimates against the six hundred thousand people living in Hiroshima and Nagasaki, our government decided bombing one, or both, of them would stop the war with a tremendous saving of lives. After explaining to the Japanese what was going to happen, the Allies requested a surrender. The answer was no, and the first bomb was dropped on Hiroshima. We waited and then asked again for their surrender after they had witnessed the death and destruction at Hiroshima. As Proverbs says, pride goeth before the fall, and the foolish Japanese leaders again said no; they didn't want to lose face, or respect, as individuals or a country. Three days later, we dropped the second bomb on Nagasaki. If the Japanese knew we only had two atomic bombs, they wouldn't have surrendered after Nagasaki, and we would have had hell to pay. Updated estimates of the final death count for both cities from the date of the bombing through sixty years of aftermath and sickness are still less than three hundred thousand.

The *Enterprise* diplomatically stayed out of the fray and quietly docked at her assigned pier. No Americans were injured during the time she was there. The *Enterprise* did have nuclear weapons, and nuclear energy, not oil, was used to operate her power plant. Naval officials today do not confirm or deny that carriers carry nuclear weapons, but they are all nuclear powered.

The Japanese political and activist atmosphere was a little calmer when we arrived, but Japan was still against any foreign nuclear power in any form on her soil or in her waters. However, the consensus is that the *Constellation*, an older conventionally powered aircraft carrier, was totally nuke-free when Pete and I were on board. We docked at the Naval Station at Sasebo under the pleasantest of circumstances in late August of '69.

The blazing lights and flashing signs of modern downtown Sasebo boldly contrasted with the dark loveliness of its older antiquated business districts. City side streets buzzed with tiny Toyota trucks unloading the day's new merchandise; the trucks squeezed in front of the small stores like piglets lying neatly in a row beside their mothers while drinking their supper. The rustic two- and three-story buildings along the narrow streets prevented any sunshine from hitting the pavement; making it a dream scene, a sight for tourists' eyes, a shadowy tube crowned with a beautiful blue sky. Pete and I didn't have the resources to see and taste all the pleasures of Sasebo, but we enjoyed walking the tiny side streets and seeing the hanging geese and ducks suspended from the gaping windowsills of the open-air markets. The small shops nestled side by side, the quietness, the cleanliness, and the courtesy of the older Japanese people enthralled us. We lustfully stared at the large prawns and mushrooms displayed in the windows of all the restaurants, but on our enlisted pay, we rarely ate anything expensive or too unusual. No menus were available in English, but each restaurant had pictures of their dishes to point at with their prices in yen beside them.

Every now and then, Pete and I would stick our heads inside a fancy downtown club and have a beer. The young bartenders were quite taken with us in our Navy-blue uniforms, and even the cabbies had nice things to say. As we sat on new-age barstools nursing a beer in one club, two young bartenders, after staring and whispering for twenty minutes, finally got up the nerve to slide down the bar.

One said to me, "You handsome boy, Joe." Another giggled and said, "You movie star; you looking good." That was nice of them, but I would have felt better had one of them been a girl.

In any event, different bartenders in different bars would occasionally give us a free beer, glass of wine, or shot of booze. They admired that gorgeous gabardine dress blue uniform I bought in San Diego that fit me like a glove. When its beautifully embroidered inner cuffs were rolled up one time, each displayed a magnificent multi-colored hand-stitched silk dragon. The cab drivers would look at us and say to me, "You ichiban skivvy honcho, Joe." Ichiban is Japanese for the best, or number one. Skivvy honcho is unabridged World War II Marine lingo that means a ladies' man. Skivvies are underwear in the military, and honcho means a hard, tough man. The title was quite a compliment. What the cabbies really appreciated about Pete and me was we were never sick in their cabs.

The restrooms at a couple of the oldest clubs we patronized had tiled floors with six or eight little platforms just big enough to step upon, each raised about an inch from the floor. When a restroom was needed, a man stepped onto two of the little platforms in the middle of the floor and carefully shot his excess beer or sake onto the lowest part of the wall in front of him. Once, in the midst of my doing just that, two young Japanese women walked in, stepped on their two little platforms, lifted their dresses around their waists and did their business. I never saw any evidence of anything but urine on any of the bathroom floors in a bar or a club. If someone had to poop, pay stalls were available in neat little public buildings outside. A lady wielding a large broom to keep the place neat and clean, but also celibate, was always there.

"Dude, that restroom freaked me out. I think Commodore Perry was a member of that club; you know he was here in the 1800s," Pete joked. "No wonder we had to pee on the floor; they've never heard of a urinal." We had taken a break from drinking long enough to go find a restaurant that served the giant tempura prawns and flame-kissed skewered octopus we had been talking about for an hour.

I marveled at the tidiness of the cities we visited in Japan. Homeless beggars, their blankets spread for the whole family, camped at night on the city sidewalks. By early morning all was picked up, rolled up, and gone. The beggars wore clean clothes and had too much personal pride to ask for money. If you quietly dropped a few yen on their blanket, they thanked you humbly with "Domo Arigado," meaning thank you very much.

The only excessive pleasure Pete and I splurged on while in Sasebo was having a steam bath administered by some lovely scrub women while we were in a hot tub. No, we weren't in a hot tub together; that wasn't the excessive pleasure. A couple of ladies helped us undress in separate wooden cubicles and led us to our individual hot tubs. Loosely wrapped in a large cloth towel hanging from my shoulders, I was escorted down a dozen massive steps of hand-hewn oak and at the bottom, I was led to a tub of steaming crystal-clear water large enough to hold two families, or maybe six or seven adults. 'Two families' was probably correct because the Japanese did everything as a family, even taking baths together...naked...with strangers. The two-family or six-adult tub sounded interesting. Maybe they rented those tubs out for private parties at night with candlelight and sake, and underwater lighting, and geisha girls playing kotos and taisho-gotos in flimsy robes.

The two ladies dedicated to me led me by each elbow, possibly because I had indulged in a drink or two. They repeatedly poured hot soothing water over my head and shoulders as I sat nipple deep in the steaming water. They were either trying to sober me up or rinse me off or shrivel me up like an old prune. Maybe they were wetting me down before soaping me up like at the car wash. The steam bath was perfect because the weather had been cool outside. One of the ladies began rubbing my back and shoulders with a soft scrub brush on the end of a long-handled stick that looked like a cutoff broomstick. My mind drifted in and out, and as I sat there, sitting naked in steaming hot water with two ladies close by, my head gently spinning from the beer and the heat, I began to think of Jane and her beautiful, long brown hair. The ladies looked at me and then each other; it was time for my massage. They let me rest like a steak right off the grill so my juices could redistribute themselves throughout my body. Then they led me up those ancient steps to a soft cloth mat on a long waist-high bench and gave me their exquisite, strictly professional massage…facedown. Every fiber of my body tingled, and I couldn't have stood on my own two feet if I had wanted.

A distant muffled commotion coming from down the hall signaled to everyone that the tranquility inside the ancient building had come to an abrupt end. The noise grew louder as a couple of matronly Japanese women, arguing in their native tongue til they were pink in the face, swarmed into the hallway. I correctly guessed they were upset with Pete, so I slowly rose from my mat and grabbed my towel, wrapped it around me and went to see what had happened. I immediately saw the problem. One Japanese lady was lying on the mat that was meant for Pete, and he was straddling her back with nothing on but a towel wrapped lightly around his waist. He was giving the lady a massage, and she was loving it, even enough to sacrifice her reputation with her co-workers. It looked like Pete couldn't go anywhere without causing trouble.

The bathhouse supervisor was yelling at the woman on the mat, but she just smiled and spit some Japanese at her. The supervisor stepped back like she'd been slapped in the face. The seven or so other women in the place gathered in the hallway and stared at the spectacle with their hands covering their mouths. They glared at me for bringing that vile creature into their midst. I stood there dumbfounded, not knowing what to do. I went to pull Pete off the little lady and dropped my towel. I tip-toed around the room in the nude, bent from the waist, stumbling as I tried to grasp the edge of the towel, my head buzzing and swirling. The little prim and proper Japanese women screamed at the sight, or possibly with delight. Their anger and indignation were equally divided between me and Pete and their colleague lying spread-eagled beneath Pete.

As hard as it was to believe, the masseuse was lying beneath Pete, fully-clothed and on her stomach, but this was not the place, or the time, for any escapades from an American skivvy honcho. The supervisor wanted to get Pete's attention, so she began yelling "MP! MP!" at the top of her lungs. Sitting there in his stupor, Pete couldn't grasp at that moment what an MP was. All the alcohol we had consumed during the day was still flowing through our veins; time hadn't removed it from our systems. The supervisor ran to the front door, opened it, and yelled "Shore Patrol! Shore Patrol!" That, as a matter of fact, did get Pete's attention. He was off the masseuse in a heartbeat, heading for his clothes, fully alert.

I ran back to my compartment and dressed as if the building were on fire. Luckily, we had paid in advance and flew to the door. We stopped to find our shoes and put them on but didn't lace them. I glanced over my shoulder as we pushed the little front door open and saw Pete's masseuse standing there with a big smile on her face. Pete and I were smiling, too; he was excited about their brief encounter, but it was too frivolous an affair to be called scandalous and entered into the pages of the criminal record of a dreaded ugly American.

10. FILIPINO CHICKEN-ON-A-STICK

A few days later the *Constellation* headed for Yankee Station right on time. It was now September of '69, and we began flight operations as soon as we hit our designated spot in the Gulf of Tonkin. Our air wing flew support sorties into South Vietnam and a few other classified places, as targets were identified and orders were issued. My fellow aviation fire control technicians (AQBs) and I knew where our jets were heading on their sorties because we entered their coordinates on the jets' computers before each flight. Our air crews were getting itchy to get into the fray once again.

The U.S.S. *Pueblo*, a small American spy ship, had been captured in January of '68 off the coast of North Korea, and the ship's eighty-three-man crew was imprisoned. Our government's official defense to the charge that we had illegally intervened into North Korean territory was that the *Pueblo* was sixteen miles off their coast in international waters. North Korea said it was only twelve miles off the coast which put the ship in their waters. This was the worst incident between the U.S. and North Korea during a two-year period of extreme crisis and tension. That December, almost a year later, after kissing some North Korean butts, our men were finally released.

A U.S. spy plane was shot down four months later by two North Korean MiG-21s with the loss of all thirty-one men aboard. Five months of pussy-footing after their deaths, someone finally found the guts to rattle North Korea's cage, and in September of 1969, the *Connie* was sent to the Sea of Japan to intimidate those bastards. Flight operations continued as we sailed to a point straddling the border between North and South Korea and stayed there a month until we were recalled. All members of the *Constellation* were awarded the Navy Meritorious Unit Commendation for "valorous and meritorious service" during her response to North Korea's hostile actions.

On a sad note, a month after we returned to Yankee Station, a transport plane carrying over twenty-five sailors from Subic Bay to the *Constellation* and another warship crashed into the sea. We never heard why it crashed. Although in a war zone, the carrier made an adequate, but hurried, search for any survivors. No bodies were recovered by our search helicopters or accompanying destroyers or other Search and Rescue teams.

Rest and relaxation in the Philippines soon gave us a big boost in the arm. After we got through a couple of nights in Olongapo City, many men also needed a big boost in the butt from our Medical Department. The *Connie* was replenished with weapons, fuel and supplies. We drank beer and listened to Filipino bands play American music in nightclubs. The singers usually lip-synced, but the best, many of whom could not speak a word of English, had learned to soulfully sing the current hit songs in English without knowing one thing about what they meant.

A typical day or night in Olongapo started with a walk across 'Shit River' at the entrance to the city. It had that horrible name because it had that horrible smell. It was basically a narrow river, or canal, of stagnant water that looked and smelled like raw sewage and separated NAS Subic Bay and Olongapo. Unfortunately, when the fleet was in, the slimehole was populated at almost all hours, but mainly in the evening, by preteen boys and girls who begged for quarters. The kids sat in the front of a couple of small boats floating in that slime and yelled up to the sailors and Marines on the bridge, "Sailor, throw me coin." When one was thrown, usually a quarter, a couple of the closest kids dove into the canal to grab it off the bottom of that dark, stinking river of filth.

"I will not throw them coins," Pete argued with me, "I'd rather see them starve to death than have to jump in that filthy stinking slime to get a damn quarter. I think the police should kick them out of the river. What kind of diseases do they have?"

"Would you rather have them all starve?" I asked.

"No, I'd rather have them eat and live and be clean and happy. I think of them like I do about wild animals in a zoo. I'd rather for them to be extinct than to live their whole lives in a stinking cage, jumping from one cinder block wall to another, never touching a blade of grass, or smelling fresh air."

"Are you comparing these kids to wild animals?" I asked.

"Yeah…Maybe I am, but not in a bad way, nothing mean; it's not their fault."

The sins of Olongapo could be ignored if a sailor chose not to look or breathe, but the conditions under which those kids lived was horrifying. Surely, they were orphans, the product of a twenty-minute relationship between a Filipino prostitute and an American sailor or Marine; a sad thought. Unfortunately, a sailor who's been at sea for thirty straight days, working twelve hours a day, every day, with no breaks, and who's looking for a place where he can get a beer and a semi-decent meal, didn't have time for such sad thoughts.

That semi-decent meal in Olongapo could just as easily be found on a sidewalk as in a restaurant. Many Filipino street vendors fired up their little hibachis right on the city sidewalks and grilled their meat as the sailors strolled by. There were no meat inspectors in Olongapo, so bon appetit and good luck!

One night, Pete wanted a piece of barbecued chicken but was worried about getting norovirus or salmonella from a street vendor. We stopped beside an old frail man in a dirty white shirt squatting beside a tiny charcoal grill, undoubtedly made in Japan. The java sauce smelled delicious, and the skewered chicken leg and thigh looked very appetizing, although its skin looked too purple-brown and leathery to me. Pete said it looked just right to him and bought a piece. The vendor had no napkins, no plates, no forks, nor knives. Pete took his first bite from the thigh while the chicken was still on the stick, eating it like a kid's corndog. The skin pulled off easily, hot and juicy, but all the skin came off in one large piece. The sauce was thick and delicious, Pete said, trying to keep the excess skin and sauce off his clean white uniform. He held the chicken with one hand and tried to tear the dripping skin into two pieces with his other fingers and his teeth. I just tried to stay out of harm's way. He wiped his mouth with the back of his hand, smearing grease and sauce over his cheek. Chicken grease slid down the skewer onto his clean hand. Pete asked the man if he had a rag or a napkin, and the old man shook his head and pointed to his ear like he was deaf. I gave Pete my clean handkerchief; apparently, I was the only sailor who ever carried one, and he handed me the skewer and the hot chicken-on-a-stick.

"What a stinking mess. This was a really great idea, Pete; just throw my handkerchief away when you're through," I said.

While Pete wiped up, I moved under a streetlight and inspected the chicken. I noticed the chicken leg had a shin bone. I told Pete none of the chicken legs I had ever eaten during my entire twenty-two years on Earth had shin bones. The legs I knew and loved had a bone in their center with meat all around it. Whatever Pete was eating was not a chicken; no, it was an animal that had done a lot of fast-paced running in its time.

"Hey, Pete, what running animal has a leg the size of a chicken's? A small dog? A house cat? Maybe a monkey?"

"I know what you're doing, man. You're trying to put me off my delicious, oriental, Filipino, sidewalk-basted chicken-on-a-stick so you can take it for yourself, but it'll never happen, Joe. So solly; so quit your scheming." We talked in imitation Japanese at times, 'Joe' and 'so solly' were our substitutes for 'man' and 'so sorry.'

"No, I really wonder what animal it is," I said.

Who knew, but in Olongapo, it was best not to ask too many questions, or get too many answers. Pete finished eating whatever it was, and we laughed about its source, but I could tell he was worried. He had been too cheap and hungry to throw the messy thing away. He needed to wash up with a little water, and I had to pee, so we ducked into a club to use their toilet. We passed through tables and chairs crowded with laughing sailors and toothless prostitutes on the way to the back, where a large white handmade sign nailed above the door said "TOILITS" in black house paint. Walking down the dark hallway, sloshing through the water from a leaky toilet, we opened the door to the men's room. The stench almost knocked me to my knees, but I really had to pee, so I held my nose and stepped up to the urinal. Pete struggled with the sight and smell. The toilets were missing their seats and lids and were full of human waste floating with pieces of used paper. The floor was littered with wet newspapers and flyers and other trash. The filthy urinals were yellow-brown with age and urine and cigarettes. Pete turned to leave but couldn't make it. He threw up his delicious Filipino chicken-on-a-stick and a whole lot of beer, three times, all over the floor.

I looked at the mess and said, "I'd clean this place up for you, Pete, but what the heck, nobody'll even notice it in five minutes. Let's get outta here before I throw up, too."

When we were in Olongapo, anything could happen. One evening I treated myself to the only steak I ate during my entire tour. Well, at least I had ordered a steak; that's what it said on the menu. It came Filipino style, covered in a thick dark gravy that was quite tasty. I pushed the gravy to the side with my fork and noticed the steak was rolled up like a cinnamon roll. The meat was one continuous strip about eight inches long, curled in a circle to look like one hunk of meat.

"What is that?" Pete asked. "Are you going to eat it?"

I wondered aloud, "What is it indeed? Monkey meat? No, the strips are too large and thick for monkey meat. Water buffalo? No, they're too old and tough to be eaten by anyone but their owners. You know they have to plow those stinking mud puddles until they're too weak to walk? Beef? I don't know. What kinda beef could this be?"

"There is a small cow, some Philippine Breed, around here," Pete said, "but if that's real beef, why is it curled up like that?"

Whatever it was, it was delicious. Many sailors complained on the ship after each port call, no matter the country, how bad the local foods and hamburgers were. The reason their burgers sucked was because the people in that country never ate burgers in the '60s, so they hadn't been cooking and perfecting them for fifty years. Go to a Krystal or your local hamburger joint in the States and ask them to bring you some pasta primavera, or linguine, or lobster Rockefeller. It'd be the same thing; it'd suck.

"These guys make me sick," Pete said. "They go completely around the world and only try American hamburgers, hotdogs, and pizza. They are miserable; they complain all the time, but most of all, they don't know they are missing an experience of a lifetime."

"I agree, and don't even try to talk them into trying something new. They just turn up their noses and say 'gross.' It's like throwing your pearls before the swine, which I think means don't waste your breath," I said.

Pete and I hired a Filipino to drive us three hours to Manila to see some jai alai played at the largest fronton in the Pacific. Our guide dropped us off out front and said he'd return in another two hours. We were the only Americans in the entire crowd of approximately four thousand Filipinos. We were dressed in starched white uniforms and stood head and shoulders above everyone else. We felt like two black guys standing in the crowd at a Klan rally or trying to get in the dining hall at Ole Miss in the early '60s. Communist rebels in the mountainous Philippine jungles were stirring up trouble, and people were being kidnapped and killed all the time. Explosions around Manila were frequently heard.

"Pete, are we in one of those places we were warned about? Maybe we should have read those posters back in the shop; I'm getting a little worried. Maybe this wasn't such a good idea."

"Relax, Smitty, have another beer. After four or five San Miguels, we won't really care. If we get captured or killed, it'll be cool. We'll be in all the headlines in the papers back home. We'll be heroes, and it'll give us an excuse to finally kill all the Flips, those Communist Flips."

Most sailors called the locals 'Flips' but never to their face, or where they could hear you. They were generally great people, but most were poor and lived in huts, and many young American boys were biased in their opinions of them because of that. The Filipinos we usually associated with were maids and prostitutes, street vendors and naval base employees, cooks and Jeepney drivers, and all those people were pleasant, peaceful, and industrious. I can never recall a single instance of a Filipino injuring an American for any reason. The worse thing that happened to me in the Philippines occurred when a ten-year-old boy smeared my dress shoes with black shoe polish. He snuck up on me out of the shadows of some bushes along a dim Olongapo sidewalk. Those street urchins usually paraded down the middle of the street, their polish and brushes in a shoebox in one hand and a rag laced with polish in the other. When they got close and a sailor was drunk or unwary, they jumped onto the sidewalk and smeared that nasty grease on the toe of his shoe. I kicked that boy's arm away with the side of my foot when he reached over and ruined my shine. I wanted him to know I really didn't appreciate it, and then I had to pay him a dollar in extortion money to re-polish them. The boys had territories, but there were three or four on every block like drug dealers or prostitutes in what used to be just the big cities.

Back on the ship after Manila, Pete and I decided to hit the beach one last time before we shoved off for Yankee Station. That night we went to a popular club called the West End. While drinking a beer or two, we naturally attracted a number of bar girls. The whole situation was hilarious; three of us sat at a table drinking beer and laughing, surrounded by scantily dressed Filipino women. Our setting was similar to the settings in all the other clubs in the whole lousy town, but some clubs were better than others; slightly better furnishings, slightly better music, slightly colder beer, slightly younger girls. Our last night of fun was interrupted by a fight when a barfly sitting at our table jumped another girl who walked in with her steady sailor. These girls took their sexual duties seriously; I didn't blame them; it was their livelihood. They built up relationships with specific sailors and expected them to come see them, and only them, when that guy was in port. Most guys whom the girls considered their 'steady dates' were enlisted black-shoe sailors from West Pac ships that frequented Po City many times.

We sailors on our first cruise had never been in these clubs before, so all the girls were new to us. For the majority of us, being around bar girls in a rowdy atmosphere where the rules were few was shocking and thrilling; when the fleet was in, Olongapo was like being in Dodge City in 1873. The sidewalks were packed with men in uniform walking from club to club, drinking and laughing and eating hibachi food. The clubs were full of loud, drunken sailors and Marines and almost as many prostitutes. Occasionally, a bar girl screamed at a sailor drinking with another girl and rushed over to scratch the girl's eyes out, but seldom his; that was bad for business. Comparing his unfaithfulness to the flitting of a butterfly from bloom to bloom, she would yell, "You no butterfly me!" She would grab an available knife and take off after him, knowing the other girls would run and stop her before she was able to do any damage. As the storm clouds cleared, the girls would huddle together and whisper about the unfaithfulness of that sorry two-timing sailor just like a bunch of high school girls in Georgia. There was usually a lot of excitement when you were in the city. Pete and I went to these clubs just for the beer, of course; the crazy lawlessness was just an interesting bonus.

Many times, a little three- or four-year-old kid ran around the club tables looking for his mother who was at work flirting with sailors. The often naked child was quickly taken away by an old mama-san because that too was bad for business; horny men, booze, prostitutes and kids didn't mix well. Those kids lived in rooms above the bars where the girls worked. Other single mothers and old women shared the kids' care. Bargirls, smiling from ear to ear, usually sauntered over to a sailor's chair and sat on his lap with an arm around his neck and asked him to buy her a drink. If he agreed, she received a small glass of iced tea dummied up to look like liquor, but the sailor was charged the full price of a mixed drink; it was like a cover charge. The fake booze was a necessity; it kept the girls sober. They had to be; no one likes a drunk, except one drunk sailor for another.

The bar girls stayed neat and clean because the competition was tough. Some had been cute, but as they got a little age on them, their health deteriorated to a point where they began to lose their teeth. It was fairly common to have a fifty-year-old woman squatting beside your table flashing you a toothless grin and asking for a beer and a cigarette. Of course, saying that most of those women had toothless grins is pretty harsh; many of them did have one or two little black stubs hanging from their upper gum, but usually not side by side. And by the way, squatting is probably not the best position to be in if you're not wearing panties and you're over fifty. But no one minded those old mama-sans; we were just one big happy family. We laughed, and we laughed.

One night after beer-drinking in town, Pete was a little tipsy but decided he needed one more beer before he called it a night. An enlisted man's club sat on the edge of Subic Bay almost a mile from the ship's pier. I was tired and sleepy and only wanted to go back to the ship. Pete was tired and sleepy but wanted to go get another beer. When you're best buddies, which we called 'asshole buddies,' you sometimes do things you don't really want to, just to keep from upsetting him. Pete and I were on foot, but I reluctantly agreed to walk with him to the club. I knew the fool needed a babysitter which, in this case, meant the blind would be leading the blind. We found a path between the wild bushes covering the sand dunes which slowly dropped fifty feet down toward the club; we could see its lights below us. We crossed an unmarked paved road leading down the hill, and Pete noticed a forklift parked on the roadside. I walked right by it without a thought.

Pete called out, "Hey, wait a minute. Let's check this out."

The forklift was the largest I'd ever seen, but Pete grabbed its ladder and climbed right up. I said, "Pete, come on down from there. You're going to get hurt. You don't have the key." I should have kept my big mouth shut, because that inspired Pete to search for a key.

"Nope, there ain't one, but there's a button here that looks like a starter."

Pete pushed the little black button, and the forklift's engine roared to life. I was standing on a road in the dark, on top of a hill, and my best friend was sitting on a gigantic forklift with the engine running. He was three-quarters drunk and so was I, but I was scared thinking of what was coming next, and he wasn't. I had that creepy feeling that something bad was going to happen. Pete put that piece of junk in gear, and it slowly started moving forward. He turned it onto the road and told me to get on. We were about a mile from nowhere, and if Pete left me, I'd be all alone in the dark, late at night, just about drunk, in the Philippines. I had no choice; I jumped onto the forklift's ladder and hung on. There was plenty of room for me on the seat, but I stood on the ladder and hugged it to my chest with both arms. I wanted to make damn sure I could jump to safety if it began to turn over.

"Come on, Pete, cut those screwing lights on before you kill us."

"Get on up here, little buddy, so you can see."

"See what, you jackass," I yelled. "I can hardly see you." Which was good. Pete was sneering crazily in the dark like he was possessed, a ghoulish smile spread wide across his face like Dr. Jekyll's chemically-produced alternative-personality Edward Hyde, spittle spraying from the corners of his wide-open mouth.

There were no lights on the forklift, or Pete chose not to cut them on. He was gaining confidence that he could drive the monster, so he increased the speed. The machine must have weighed ten tons, and it was perched on top of those little bitty wheels. I knew we would turn over and die. We went faster and faster, and I was scared to death. But the club was getting closer. We leveled off at the edge of the parking lot surrounding the club, and I felt a little better. The downhill trip was horrifying, but we had made it alive. I calmed down and began to think that maybe Pete was right; maybe we were smart taking the forklift to the club. It saved us a lot of energy that could be used to lift more beer bottles.

A sailor on watch with some kind of night-vision infrared device noticed us when we first began messing with the forklift and immediately alerted base security. The Shore Patrol operated like bouncers for the club, so they just walked outside into the parking lot. They couldn't see us in the dark but could hear our mighty racket as we flew down the hill. Our yelling and screaming, and the forklift's clanking and squealing, could have been heard by a deaf man. They were waiting for us to get there.

We were still cruising along, but halfway across the parking lot, with only the lights of the club before us, a searchlight flashed on, blinding us. Four more lights came on; two jeeps had turned their high-beams on us. We rolled to a stop as two sailors on Shore Patrol surrounded us.

"You guys are in a lot of trouble," said the First Class Petty Officer who was in charge. "You stole this forklift, and it will be hell to pay. Your asses are done."

Pete was unfazed with it all because he was actually drunk at this time. Sometimes he was unfazed just because he was stupid as hell, but I was cold sober at this moment. It occurred to me that these SPs were taking us to the brig, that we were going to be court-martialed, and we were not going to like it. Friend or no friend, I began believing I could get into enough trouble on my own, and maybe I'd be better off without ole Peter Boy.

Authorities dealing with problems concerning sailors on liberty identify them and call their unit. Chief Cranston came to take us into custody. He gave us a hard time in front of the SPs and drove us to the ship. He said we wouldn't know our fate until he had talked to the Captain.

The Captain? The Captain of the ship?

My head was buzzing from the booze, and I was young and stupid and drunk enough to actually worry about the Captain of an aircraft carrier taking the time and energy to review a little misdemeanor like ours. I knew we were going to get punished but didn't know how badly. As it turned out, our Chief Warrant Officer made us miss two days of liberty at our next port, the beautiful city of Hong Kong. After all, he knew boys will be boys.

11. THE PIGEON AND GHOST FRONTS

We returned to bombing Vietnam at one place or another in late November of '69, but after a month on line we sailed to Hong Kong. Most ports of call had piers that were large enough to accommodate gigantic military ships, but not Hong Kong; the harbor was old and its land too valuable to reconfigure for more modern ships. Besides, the Communist Chinese wouldn't have liked it. Dropping anchor in that magnificent harbor or in any port made it a little inconvenient to go ashore. Instead of just walking down the ship's gangplank into the bustle on the pier, sailors had to putter along in small boats or ship's liberty launches into port. Depending on the weather and the time of day, sailors waited with twenty or thirty other guys on a little deck or platform below the hangar deck, or on the fantail at the very rear of the ship, looking out into the sea, until a Navy or commercial vessel came alongside.

Loading and unloading a boat was easy unless the weather was bad and the sea was choppy. Hong Kong was so naturally sheltered that choppy seas was seldom a problem. In bad weather, sailors didn't worry about getting wet; they worried about getting onboard the launches. The major problem was they stood on a steady, barely moving ship and stepped onto a boat bobbing up and down twenty or more feet with every swell. It wasn't hard, but many of the ship's crew were uncoordinated boys who apparently had never been on a boat or ferry before the Navy. They didn't have the ability to time the ascent of the rising boat to its crest and then just step onto it. A carrier has many decks and exits in which to leave or enter the ship. For ease and safety, the deck which was selected for the exit point for going ashore was determined by the height of the sea swells and where the smaller boats would crest.

Going into port in bad weather was troubling but coming back in the same crap had even more drawbacks. The liberty boats were not little things, they carried thirty or more men at a time, and their tops were covered with a heavy tarp to keep the blazing sun or the cold rain off sailors. In bad weather, the boat's crew would pull more tarps down around the sides, so the only portion of the boat exposed to the elements was the one entrance, or exit, hole on the starboard side near the bow. The tarps around the sides were a godsend, but Pete and I learned early in our naval careers that the smartest way to travel from ship to shore in bad weather was at the boat's open exit where there was a cutout in the tarp. These boats had no seats or chairs, just long wooden benches where a drunk could be pushed out of the way to make some room. Pete and I deliberately timed our entrance into the boats based on the last available room on the benches nearest the opening. We were usually the last ones on, on purpose.

Think of thirty sailors packed like sardines in a boat at night after three or four hours of moderate to heavy drinking. Think of a constant downpour and large waves with white tops popping all around. Think of that boat filled with sailors going up and down, up and down. Think of all those bellies full of warm beer going up and down, up and down. Think of what happened when the first of those thirty drunken sailors didn't reach the exit and threw up in the middle of that boat, and the rest of the boys started smelling that hot putrid liquid that had spewed out into the boat and onto their clothes. Think of how many other boys quickly got sick from the smell and sound of their shipmates, squeezed into those tight quarters, throwing up at their feet in a storm at sea.

To avoid that horrific mess, Pete and I learned to live with the cold rain hitting us in our faces as we cruised back to the ship with our heads stuck nearly out the exit hole. On occasion, a sailor forced his way through the crowd to the exit before he became sick, and then we made sure we got upwind toward the bow of the ship and out of his way. When a sailor threw up at the opening, that disgusting stuff headed into the wind which immediately blew most of it right back into the boat. Anyone downwind of the exit received the boy's spray all over them. The poor sailor then wanted to get out of the rain, and we pulled him out and moved right back into the opening, diligently looking for the next guy.

As our punishment for the forklift incident, Pete and I missed liberty on the first two days in port. It didn't matter; we had four other days to see the sights and only had money for three. Money was always a problem during our entire four years, even when we got tax-free extra combat pay. Tax-free was a joke to us; we never had enough income to even file a tax return.

Our shop Chief was a top-notch American sailor, Master Chief Charlie Cranston, a fine example of a man. Intelligent, witty and steady, the cigar-smoking veteran of twenty-nine years would do anything for his men. The Chief had been on liberty so many times in so many ports that liberty in an exotic place was like going to the mall to him. After six or seven times in Hong Kong, sailors got used to the scenery and the shops and spent their time having a nice meal and drinking a couple of beers just to get away from the daily routine of the ship. So it was no surprise that the Chief sat in our shop most of the time in Hong Kong. When Pete and I had to stay on the ship, it also meant we had to work on our planes and stay around the AQ shop. It was easier talking to the Chief than working.

Charlie Cranston was a great talker; he knew a ton of Navy stories. At first Pete and I would engage him in conversation just so we could stay inside our quiet, cool shop and loaf. It was quiet because the few guys hanging around were usually tired and queasy from the previous night's liberty and only wanted to mind their own business, drink a cup of coffee and smoke a cigarette. Usually in port, there were several British newspapers and some Navy magazines lying around that someone had found, or stolen, or bought on liberty. When Charlie got started on a story, either a personal story or a national or worldly one, there was no telling where the conversation might go.

Looking at the duty roster our first day in Hong Kong after everyone, except us, had gone on liberty, Pete asked Charlie, "Chief, how long you been in our squadron?"

"Oh, God, I guess it's been ... let's see ... born in '20, joined in '40, on the *Pigeon* til '48, aviation schools and teaching til '62, then here ... so it's been almost eight years that I've been here."

"Wait, Chief, you've been in the Navy twenty-nine years? You were in the fleet during WW2 ... on what? The *Pigeon*? Was that a ship?"

"Hey, man, don't make fun of my ship. You can get your ass whipped for laughing at an old salt's combat vessel, especially during the Big One."

"Sorry, you know I was just messin' with you, Chief," Pete said as he reached for a cigarette. "Tell us about it."

"Okay. Well, I was a young stud like you guys. It was June of '40, and the Germans had conquered France. It looked like we were going to have to save the French again, just like we did in the First World War. I wish some other country besides ours would save the rest of the lousy world sometimes. Mama's brother was killed in Belgium so my family hoped to God war wouldn't come again. They were small-minded people who didn't care about the rest of the world; they only wanted

'Peace at Home' at their house. They were small-minded country people who believed that every country should fight its own wars.

"I turned twenty and was just doing odd jobs 'cause it was hard to get a job back then. I said the heck with it and joined the Navy. It hardly paid a thing, but you got three squares a day and a roof overhead. I didn't have anything to lose. After boot camp and some schools, I was stationed on the U.S.S. *Pigeon*."

"Sounds like a hospital ship, Chief, like the U.S.S. *Dove* or some peace ship," I said, making up the ship's name.

"Yeah, I know it doesn't sound too ferocious, but we did our share. Immediately after the Japs attacked Pearl Harbor, Roosevelt called Nimitz at home and told him he would now be the Commander of the Navy's Pacific Fleet. Roosevelt said it was a day that would live in infamy; the attack was one of the most cowardly acts of all wars.

"Nimitz landed at Pearl Harbor and said there was such a spirit of despair and defeat you would have thought the Japs had already won. Nimitz was given a tour of the harbor with its ruined fleet and said the Japanese had made three big mistakes. The first was attacking on Sunday morning. Nine out of ten of the crewmen of those ships were ashore on leave. If those same ships had been lured out to sea and sunk, we would have lost thirty-eight thousand men instead of thirty-eight hundred."

"Wow," Pete said, "I never thought of that."

"Of course you haven't; you're a dumbass. Mistake number two was the Japs saw all those beautiful battleships lined in a row and got carried away, leaving our dry docks untouched. If they had destroyed the dry docks, we would have had to tow every ship back to San Diego to be repaired. They would have had three more years to conquer all of eastern Asia.

"The third mistake was in not destroying our fuel supplies. Every drop of our fuel in the Pacific was on top of the ground in storage tanks only five miles from Pearl Harbor. One single attack plane could have strafed those defenseless storage tanks and destroyed all our fuel."

"It's hard to think anybody could have been that optimistic over Pearl Harbor," I said.

"That's exactly what Nimitz was; an optimist. He was the right man for the right job. He was Roosevelt's greatest pick of World War II. My old skipper on the *Pigeon* was an optimist, too. His name was Lieutenant Richard Hayes. After graduating from Georgia, he considered playing baseball or becoming an attorney, but he chose a military career. He became the XO of the *Falcon* where he received his first Navy Cross for salvaging two sunken submarines. He was made an Ensign by Congress for that."

"Well, how 'bout those Dawgs," Pete said, looking at me. "You don't hear of any Yellow Jackets doing that."

"You never heard of any Bulldogs doing it either, til now." I said.

"You boys want to hear some more? You gonna listen or I'm going to read my book?"

"Sure, Chief, sure. Sorry."

"Hayes took command of the *Pigeon* in 1940 as a Lieutenant at the same time I was on her. In December of '41, we were docked at the Cavite Navy Yard in Manila Bay having our steering gear repaired. Because the Pearl Harbor attack was only three days earlier, the LT kept everybody aboard and our engines ready to get underway. We bitched and moaned and groaned, but Hayes didn't care. He knew what was coming, even if no one else did. He told his Chief, 'The damn naval yard isn't even on alert; what kind of a navy are these men running?'

"When the Japanese attacked the docks, we put out to sea. The other ships, the sitting ducks, were tied to their berths. Looking back, the LT saw that the *Seadragon* was surrounded by fire. In the middle of all that bombing and strafing, he sent us back to the docks, and we pulled that helpless sub stern first from her berth. Both of our ships were blistered, and the brim of Hayes' hat melted. He received his second Navy Cross for that."

"Man, that's terrific," Pete said. "Did the crew get a medal or something?"

"Yeah, we got a Presidential Unit Citation. The *Pigeon* was awarded the war's first Citation, and the *Seadragon* earned eleven battle stars. Immediately after the attack, we mounted two 3-inch guns and twelve .50 caliber machine guns on her; those were my babies. We called her a 'gunboat' and received a second Citation for shooting down some planes and blasting some Japs to hell. We were the only surface warship to win two Presidential Unit Citations in World War II. The LT was promoted to Captain in '45 and to Rear Admiral in '52."

"Sounds like a helluva man, and you guys were great," Pete said. "You were there, Chief; you knew what happened. Who started the war anyway?"

"The Germans started it. It was supposed to be about jobs and food, but it was about land-grabbing and empire building. Of course, many industrialists were looking for money and profits and riches. But wars are caused by deeper influences than those sons-of-bitches. Germanism was the cause of World War II. They set out to conquer and destroy the civilized world because they weren't civilized themselves. Somebody tell me one good thing that one single Nazi or Jap or Wop did between 1930 and 1948."

"That's pretty heavy. You're like some philosopher or something," Pete said.

"Sorry, guys, I get worked up about the war. I saw a lot of buddies die and for what? Hey, let's take a break; it's chow time."

There was no need to lock the shop; anyone who wanted a computer or its drum or a jet radar could help themselves. The Chief had his own mess; he couldn't eat with peons like us. That would be like fraternizing with the enemy or something.

When we returned, the Chief was sitting at our only desk with his feet on top, smoking a big old stogie. "Hey, boys, how was chow?"

"Fine, Chief, burgers and fries and ice cold milk. God, I love that milk. It's the best thing about port," Pete said. "You feel like talking some more, or do we have to go to work?"

"Pull up a chair and fire away. Everything can be done tomorrow. I just lit this big boy, and I'm smoking it all unless we go to General Quarters!"

"General Quarters. Chief...did you see that boy's funeral after we left Hawaii?"

"Yeah, that burial at sea. I saw you guys. It's something, isn't it? Saw more than a dozen on the *Pigeon* in the big war. About the third I've seen in this one. This one really was special, but the best military funeral I ever saw was my best friend's from high school."

"Fire away, Chief."

"My buddy Bob Johnson left New York in '44 with the 9th Armored Engineers and landed in France. Their first line duty was on a quiet section of the Belgium-Luxembourg-German frontier. They had never seen combat and were put in the Ardennes where no fighting was expected; that's called a Ghost Front. It was eighty-five miles long and manned by six divisions; three of them were totally exhausted from combat and were sent there for rest. The other three, including the 9th, were there to get a tiny taste of combat; they could get a bloody nose in some small clashes and see what combat was all about."

"Never heard of that, Chief."

"It's true. Bob and seventy-five thousand other Americans were on the Ghost Front with absolutely no clue the Germans were going to attack. The Germans were in place; twenty-seven divisions of well-armed elite troops and the best military equipment in the world."

"We had seventy-five thousand men; how many did the Nazis have?"

"Twenty-seven divisions, including reserves. Ten divisions of panzers and seventeen of hardened veterans. Two hundred and fifty thousand Germans. On the 16th of December, the assault began. Our men were sleeping in their bunkers and foxholes. The whole front wilted under the barrage of their howitzers and field pieces. The whoop, whoop of their mortars. The hissing of the rockets from their self-propelled vehicles. The shells from the German railroad guns screaming overhead as a thousand Nazi tanks roared to life.

"The walkie-talkie rang, startling Bob's telephone man. Bob was ordered to take his men and six tank destroyers north to plug the hole that the Germans made when they charged through the front. Ten miles from Bastogne, Bob's destroyers plunged into the tail end of the German 5th, destroying two Panzers. The only thing preventing the 5th from charging into Bastogne were the five thousand men of our 110th Regiment, which was being slaughtered."

The Chief was excitedly waving his outstretched arms and hands, illustrating his story.

"The Nazis were getting anxious. They said what's holding up our attack over there. Their problem was the northern part of a strategic road was being defended by about three hundred cooks and clerks forced into combat when the 110th fell. The men were terrified, but they never left their posts. They were given rifles and pistols and were told to hold an old castle, the place where they had been working, for two days. If they surrendered, our men on the eastern edge of town would be cut off from the rear.

"Sixteen of our Sherman tanks entered Clervaux just as my friend Bob got there. It was a classic case of too little, too late. Five of the American tanks were sent to reinforce a strongpoint in danger of collapsing and were destroyed. The remainder of the armored vehicles and tanks were expected to stop the German attack, but all of our strongholds were wiped out, allowing thirty German tanks to move into town. The Panzers knocked out nine of our tanks and three of Bob's tank destroyers. Bob was badly wounded but would not leave the field. His orders were to hold at all costs. His commander said there would be no retreat, 'you've gotta fight, Lieutenant.'

"Bob only had two Shermans and three tank destroyers left and two hundred engineers, cooks, clerks, and bakers. No one else was commissioned, so he had taken charge. Four Panzers clanked their way across town and blew his two Shermans to pieces. The Shermans shot first, and their shots were on target, but their shells had just bounced off the Panzers' heavy armor. One of the tank destroyers blasted the front entirely off another Panzer. That was the difference between a tank and a tank destroyer. Before another shot could be fired, the other Panzers, over-estimating Bob's strength, had quickly backed off.

"As darkness approached, Bob's crew was ordered to a roadblock called Task Force Rose set up five miles west. Sixty of the men were from the castle. He turned them over to the first Captain he met. Told to report to a medic to get his wounds bandaged, Bob refused, saying he would not leave. The Captain was called away, and Bob was in command again; there was no one else."

"He was some kind of man," Pete said.

"Yeah, he's what's needed to win a war; someone who knows he's going to die but keeps obeying his orders anyway. Around three the next day, Task Force Rose was overrun by the 2nd Panzers, and the few survivors, including Bob, fought a retreating action back to the lines of a second roadblock called Task Force Harper. The tank destroyers had been demolished by the superior numbers of the 2nd Panzers.

"The sun was setting, which was fine with the Germans. Panzers and Tigers preferred to fight at night; the darkness made their giant bodies harder to see. In the closing light, the 2nd Panzers struck with full force and crushed the little roadblock. The surviving men scrambled into the few vehicles at hand and tore off toward Bastogne. Halfway there, they set up a skirmish line at Longvilly, five miles from Bastogne. They were the last remaining American troops in front of Bastogne, and to their rear were the last American artillery pieces between Bastogne and two German Panzer Divisions.

"The 2nd Panzers turned and were replaced by the Panzer Lehr Division, which was heading straight for our boys. Bob coordinated the defenses in front of the Longvilly skirmish line. Tired, cold, and hungry, and wounded, he was in front of our line, throwing together a defensive wall of wagons and doors to protect his men. No one heard the bark of the lone German rifle in the midst of the mind-boggling roar of the war. The sniper's bullet went through his head just in front of his left ear. He didn't even know what country he was in, but he was in Belgium…in December of '44. As he had been ordered, he did not come back."

"Man, that's tough, Chief. He really was a hero," I said.

"Yeah. He was buried in a military cemetery in France. Four thousand other Americans joined him. In '53 his body was exhumed and shipped to Marietta's National Cemetery, and I was there. That was the best funeral I've ever seen. His coffin was draped with a beautiful flag. His pallbearers were a military honor-guard gathered around his casket. At his head stood three sailors at attention, wearing their dress blue tops and bell-bottomed pants. At his feet stood four soldiers also at attention. They were in their dress khakis. Everyone's shoes were spit-shined. Behind his casket were two Marines standing at attention, and between them was an Army color-bearer holding aloft an American flag. Behind them stood other officers. A soldier standing by the casket held his trumpet at his side. The government really knows how to put on a show. All it costs the attendees is to supply the exhibit, the corpse, which is always the dearly beloved, the sacrificial lamb."

"That's pretty powerful. Sounds great, Chief."

"The preacher said a prayer, and Bob was lowered into his grave.

From the edge of the tree line, three volleys roared across the quiet cemetery, and then the bugler played a long, slow Taps. Like I said, that was the best funeral I've ever seen, but it may be because he was my old friend."

Pete and I just sat there a moment enjoying what we had heard. The Chief dragged his feet off the desk and stood up. He stretched and said, "You boys just gonna sit here all day? Get your butts out there and find something constructive to do."

"How do you know all that about him, Chief? Is all that true?" I asked.

"I talked to the men who survived Bastogne and made it through the war; the guys who were with him from the start. They told me the whole thing four or five different times…at their reunions. I was always invited because I represented Bob to them. I've repeated their stories a hundred times. Old salts like to hear good stories like that. It makes 'em proud to be a serviceman. They love their country and their heroes."

12. BURIALS AND BOMBS

I was anxious to see the Chief the next morning; I had a lot of questions. Pete and I strolled into the shop, and there he was, sitting at his desk, going over some paperwork. He looked up and said, "Grab a cup of coffee and look at this list of things for this morning."

In port, we only did light stuff. Few crewmen worked on planes when most of the ship's company was on liberty. There was no one to help if something bad happened. "Pete, you need to replace the APU in 503. Get Smitty to help you. The B/N (the A6's Bombardier/Navigator) said it was intermittent."

"Dang, it's too early for that, isn't it? I just got up," Pete said.

"Tough shit, old buddy, get it done. It won't take you an hour." We talked like that in the Navy, many of us did, but I guess the nice boys didn't. I wasn't around them enough to be sure. "Smitty, you check out that CCU in 516. You and Pete may have to take it out. Okay, guys, finish that coffee and get to work. That's all we have today unless something else walks in. Oh, somebody's gotta replace that search radar in 507 tomorrow."

I told Pete as we headed to the hangar, "Every time he says 'somebody's got to do something,' he means me."

It was eleven thirty before we knocked out 503 and then replaced the CCU in 516. It's tight working in the cockpit on an A6. The most important thing is to not blow something up. The Chief had gone to chow when we got finished, so we did, too. He got back first.

"Chief, where'd you go? What if we'd needed you?" Pete said as we walked in.

"I trust you boys. Besides, if you needed me while I took thirty minutes to get something to eat, you'd be dumber than I think you are. So how'd it go? Everything signed off? No problems? Did you cut the systems on after you finished to see if they were up? Speak. Tell me. Write it up."

"Sure, we did all that. They're both fine."

"Great! I think I'll put you two idiots in for a damn medal," the Chief joked.

"You're so funny; you should have your own show. But we liked your stories about WW2 yesterday. Why does the government put so much time and money into funerals?" I asked.

"You want another history lesson, is that it? Okay, sit your butts down, and I'll tell you. We got nothing else to do. Y'all can clear out all this trash and fill that coffee pot when we're through. That's all we have, and I'm not going ashore until tonight."

"Got a date tonight? Some cute little white Russian Hong Kong mama-san?" Pete asked.

"You know better than that; I'm married. And too old for that crap. You do know those white Russians are actually Belarusians, not real Russians. That's all they are. Bela means white in Russian, so white Russian just means Bela Russian. But you're right; they've got the best looking women in the world."

"Thanks for the lesson, Albert Einstein," Pete said. "So, whatcha gonna do? Find a couple of old ladies and go shopping?"

"Hell, no. I just want some Peking duck and those big red prawns smothered in Cantonese sauce so hot it'll make my butt burn for three days. I'll have to get a stick to kill the tigers as they claw their way out. And some ice cold Chinese beer."

"Sounds good, but don't rub it in; you know we can't go. But why do we spend so much money on funerals?" I asked again.

"Okay. Let's see...In the 5th Century B.C., the Athenians and Spartans cremated their dead on the battlefield out of respect. They brought their ashes back home, and everybody brought flowers and ornaments to put on their remains, so they planted them in a garden. They honored the boys who died protecting their nation; a nation which was only what those men made of her. It was a special tribute. But that was forgotten. As late as the American Revolution, soldiers were left to rot in the field, never buried, with or without honors. But photographers began selling pictures of the dead soldiers on the battlefields during the Civil War, and people saw war as it really was. They saw their sons puffed up like giant balloons ready to pop, and after they popped and the gases blew out, they saw their boys' dried, broken, scattered bones picked over by buzzards and wild dogs."

"That was Mathew Brady. The photographer," I said.

"Yeah, I think it was. But that was some gross stuff. People had never seen it before; it shocked the hell out of them. Hey, see if there's some coffee left. Fix me a cup."

"Your wish is my..."

"Anyway, the burial details, their faces covered with rags dipped in cheap perfume to cut the smell, carried hundreds of thousands of rotting bodies to holes dug in the ground. They stomped on arms and legs stiffened by rigor mortis until the body was crammed down in it. Soldiers saw how bad they were treated, just covered with loose dirt like a farmer planting a crop, and wrote to their mamas begging them to make the army send 'em home if they died.

"Abe Lincoln heard the crying and set up fourteen National Cemeteries for dead Federal soldiers; they've got over a hundred and fifteen now. They spent five years digging up and reburying three hundred thousand Federals in Northern cemeteries near the southern border.

That wasn't too hard 'cause, in the beginning, most of 'em fell on the battlefields in the Border States. The Confederate dead lay just as they had died; nobody touched a single one. Some southern women organized a memorial society to find and bury their dead sons at their own expense."

"That sucks," Pete said. "My Confederate ancestors died in Pennsylvania and West Virginia. So did Smitty's. What happened to them?"

"I just told you that, dumbass; they rotted in a field. Some hungry wolves took their heads away. I don't want to get into Civil War rights and wrongs. Not now. Anyway, around 1917, a system to send our World War I dead home was set up, but our generals said 'no way;' it would cost too much money and time to dig 'em up and ship 'em home. They were against it because the money would come out of their budgets and the workers out of their ranks."

"That's disgusting," Pete said.

"You know how the military is; they never want to reduce their budget from one year to the next. They're afraid it won't be raised again when they need it. Money out of their budget for reburials meant less money to spend on tanks and cannon and rifles. It's hard to get more money unless there's a war going on, and theirs was over. Hell, I saw sailors at the fantail on the *America* last year throwing tens of thousands of dollars' worth of unused brass parts in the ocean just so they could requisition them again and keep their budgets up. You ever heard that? There's some sick crap in this man's Navy, but I'm sure it happens everywhere. I tell you; it's pretty hard to take," the Chief said.

"Well, our allies were also horrified; if we dug up our seventy thousand dead, the British would want their seven hundred thousand boys dug up, too. And France said if that happened, they'd have nothing but trainloads of decomposing bodies crossing the country every day, so they forbade the removal of any bodies."

"Yeah, that sounds about right coming from those pansies. They couldn't wipe their asses without us," Pete said.

"You got that right. Surveys were sent to eighty thousand American families asking if they wanted their boys returned home. Their boys could be dug up and returned home, or they could be buried in the new American cemeteries in Europe. France lifted its ban, and we spent two years and over thirty million dollars taking our men from their plain battlefield graves. But make no mistake about it; those precious graves of our dead boys were as full of honor and glory as any hallowed grave that ever held a human. At the end, forty-six thousand soldiers were returned home, and thirty thousand were reburied in Europe."

"Thirty million dollars. That's a lot of bread. About the same as a new A6," Pete said. "That's what they cost in '67, but I heard they've gone up five million in just the last two years."

"Yeah, that's right, I've seen the numbers…But in 1919, some Americans were opposed to the reburials. They said their sons' bodies would be horrible things, so broken and decomposed their own mothers would collapse if they saw them. Some Senators, even your man Tom Watson, voted against the Tomb of the Unknown Soldier, saying there was no telling what was being sent back to us to bury in that sacred shrine. He said the bodies could be French or German; think of that, we could have a national war monument with a stinking German inside. Some Europeans took offense that we didn't want our boys left in their country. They said America feels it's morally superior to Europe, and American soil is God's favorite country, and the rest of the world is just dirt.

"So you can see why I wouldn't have missed Bob's re-interment for anything. I had to take six days leave to go to Atlanta and back to Naples to do it. But it was worth it; he had been my best buddy, and he was a hero."

"Good for you, Chief, you make it sound so real. How do you remember all that stuff? What's it take to be a hero, anyway?" Pete asked.

"Well, I'm at sea about two-thirds of each year, and I read whenever

I'm not working, instead of going out and getting drunk every damn day like you do. I remember it because I'll read maybe fifteen books this cruise, what's that...about one every two weeks, usually history, mostly American history.

"Listen, boys, heroes...there are many heroes out there; most are just like us. There are good and bad people everywhere. There are strong and weak people, but we seem to have something in our blood that makes us react to danger. The majority of Americans don't want to hurt anybody, but we're ready to stand up for the right thing, most of the time. That's what makes a hero; someone who stands up and reacts to danger when it's necessary. But most people are clueless and helpless, and there are evil men who feed on those people without mercy. There are hordes of wicked people out there, and the moment you pretend that's not true, you become a weak person living in denial who fantasizes that there is no evil in the world.

"There is no safety except by believing that evil is possible in all men. But if you say there is no evil, and you want to live in denial, then live in denial, but you'll pay the price. When evil strikes, you and your family will die if no policemen...our greatest heroes...or no personal weapons are nearby to save them. We are not animals, boys; we can choose to be heroes or whatever else we want. My friend Bob chose to be a hero and did what he had to do, at the cost of his life, to protect our country and his family from evil. I have done the same thing for my country and family for almost thirty years, and so will you when the time comes. I want you to stand up and act like heroes when you are given the chance.

"Now haul all this trash out and go get me a pot of coffee. Then get the hell out of here and go find someplace else to waste your time. I've got a paper to read and some reports to finish, and I might start a new book. I can't hold your hands today."

Pete and I chugged our coffee and did what he said. We didn't close the shop door when we left because it didn't have one. The shop was on the hangar deck and had a large portal to step through to enter. I went below and sacked out a couple of hours, got up, found Pete and went to chow. It was quiet; everyone was ashore. "Let's go watch Hee-Haw," Pete said.

The ship had two TV stations; one always showed live action on the flight deck, and the other showed old movies and TV shows like Laugh-in and Carol Burnett and Glen Campbell that had aired four or five months earlier back home. The screen was about two feet square and black and gray. Chief Cranston was sitting there when we walked in.

"You boys back? I thought you were going into Hong Kong. Oh, my mistake, I musta forgot; you're restricted for that little forklift stunt you pulled."

"Very funny. Har-har! I thought you were supposed to be kind to us like a father figure."

"Hey. That's the Boy Scouts, not the frigging Navy. Or maybe the Air Force. And believe me; you wouldn't want me to be like my old man was to me. That was one mean son-of-a-bitch," the Chief said.

"You not going for your Peking duck and giant prawns and those mushrooms bigger than a plate? Maybe you can roast one of those ducks hanging by their necks at a street stall."

"Now, how in the hell could I roast a damn duck? I forgot to bring my hibachi," the Chief said.

"You could pay one of those coochie-coochie girls to cook it for you at her place."

"Yeah, but I'm not as sick as you guys. What're you doing in here anyway?"

"What else is there to do? We want to watch some TV. Hee-haw's on soon. Or you can tell us some more about when you were in the big war."

"Oh. You want some more blood and guts?"

"You're a Christian; tell us how you felt about the A bomb. You were in the Navy when it was dropped," I said.

"Well, alright, I can do it…It was 1945. I wrote my mama while we were cruising in the Med. We'd been off the coast of Italy and Sicily for a month making sure all the Axis hotspots were out. Our boys were fighting inland, and we were doing what we could to supply and support them. Y'all know the Axis? The Germans, the Italians, and the Japs. We heard Roosevelt had died, and my mama loved his ass. I told her something like 'our country has suffered an awful blow. We've lost our dear president. I remember how much you loved him. It must have been a great shock to you. If we have ever needed a great man at its head, our country needs one now. I don't think a new president can reach the heights that Roosevelt did. I'm sure that some other people will think otherwise.' I wanted her to know I felt her pain, so I said that Roosevelt crap just to make her feel good; I knew Truman could take his place in a heartbeat.

"Roosevelt died in Warm Springs; I guess you know that since it's in Georgia. Vice President Truman was an unknown quantity to most; Roosevelt had never let him do anything, and many people were scared to death. He used to sell clothes out in Missouri somewhere. The Nazis were still fighting in Europe, and an invasion of Japan was staring us right in the face. Henry Stimson said that maybe eight hundred thousand of our boys would die in that invasion. Our casualties killed and wounded would be between two and four million. The Japs were fanatics and would fight to their deaths with no surrender. Estimates were that four to ten million Japs, including civilians, would be casualties.

"But Roosevelt's death was no reason for us to have had any fear about Truman. He had no trouble deciding to use the atom bomb on the people who had helped start the war. The alternative would have been to condemn a million of our boys to die fighting them. He was a religious man and knew the Christian Just War Theory.

"Truman had a strong character; he was going to do the right thing. When Roosevelt died, Truman started rereading the best war book ever written. Page after page, scene after scene, thousands of soldiers and non-combatants, including old men, pregnant women, and children, are slaughtered and disemboweled, and their heads and fingers and toes are cut off. Tens of thousands of people are killed like that on almost every page. Their animals are destroyed, their babies are slammed on walls, and their cities are burned to the ground. Their wells are filled with salt. And that was all in just the first half of the book.

"Truman read that book day after day, wondering if all those deaths were justified. He believed in using war and its devastation to somehow bring justice to the world. He often thought about the concept of divine justice where evil is destroyed so that good can prevail. He wondered if he should kill half a million people to save four or five million or more."

"Chief," Pete said, "what book is that? The Koran?"

"No, dumbass, I doubt that Truman ever read the Koran. But you can read those stories for yourself. It's a book you probably grew up with at home, the Old Testament. Truman read it every day. He worried he was making our men break one of the most important of the Ten Commandments by killing our enemies while defending the civilized world. Not at random or for no reason, but during a war. Does 'Thou shall not kill' refer to the killing of all people, or just innocent people and not warriors? Are there times when only death will accomplish divine justice? Truman wondered if he could live with the Just War Theory."

"What was that? A just war? Are any of them just? Did Truman make that up?" I asked.

"Well, no, the Catholic Church originated the theory and not for

World War I or World War II. St. Augustine made it up in the 300s, and Thomas Aquinas refined it in the 1200s. Most Christians believe in it...There must be the advancement of good and the destruction of evil to justify deadly force. Its object must be a complete peace between the warring countries, and bloodshed without any hope of victory can't be justified. What else? Non-combatants and prisoners can't be intentionally harmed, and a just war mustn't cause more harm than the evil committed by the enemy.

"I'm a Methodist, and our Book of Discipline preaches it. Most Christians realize when peaceful alternatives have failed, war is preferable to unchecked aggression and tyranny. My Church recognizes that sometimes there must be righteous warfare, but we deplore war. What do you think? You think stopping someone from killing more millions of innocent people after they've already killed thirteen million in death camps is a just thing? Can you think of one real person, some realist who doesn't think like a saint, who believes all those Nazis we killed didn't deserve to die? The Italians and the Japs killed almost as many civilians as the Germans, and they got what they had coming, too."

"I agree," I said.

"In August of '45, Paul Tibbetts and his B-29 gave the Japs some divine justice when he dropped our Little Boy atom bomb in their laps. After three days, the Japs still refused to surrender, and we dropped another one on Nagasaki, and then they surrendered. You already knew all that, but many people criticized Truman for using those bombs on so many civilians. Truman sympathized with them but said the only path that he could follow was, 'When you have to deal with a beast, you have to treat him as a beast. It is most regrettable, but true.' Wow, what wisdom; when you deal with a monster, treat him like a monster. Harry Truman, a God-fearing man, saved an estimated ten million deaths and horrible injuries by using the A bombs; a million would have been our boys. I was on my way to Japan when it happened; I might have been one. Okay…You guys had enough? Nothing else to say? What you want to do?"

"One last thing, Chief. How about Hitler. What did you think about him?"

"That bastard? The 'Little Corporal' they called him because he was a corporal in the German army in World War I. He was about as impressive as you losers. My favorite name for him is 'the man most likely to be in Hell.' That's what my grandma used to say. Hitler saw us speeding across France and knew something had to be done or his war was lost. He decided that the best defense was a good offense, so he made up a plan where his army would punch through the Ghost Front where my friend Bob was and go on to capture Antwerp. He thought that would destroy thirty Allied divisions and give him a port near the North Sea to resupply his forces. His plan only needed two things: our complete surprise and enough bad weather to ground our planes. He actually believed it would work.

"My problem with Hitler is this: when he killed himself, did he go to Heaven or to Hell? I don't know, but maybe he took Jesus as his Savior on his deathbed and was forgiven for the deaths of those thirteen million innocents in Europe. Maybe his sin in starting World War II was brushed aside on Judgment Day. That's what bothers me. As a Christian, I'm supposed to forgive him and hope he repented and went to Heaven, but I can't. I'll go to Hell myself before I forgive that monster. No one should; it's God's decision to forgive unholy murderers, not mine. And after Hitler's death, His Majesty the Emperor Hirohito of Japan surrendered and was demoted to a living, breathing, effeminate little governmental puppet. It turns out he was not a god, after all."

"Damn, Chief, you've still got fire inside those old bones of yours. Any more stories?"

"Not really. You've worn me out, but…there is one more story. You'll like it because it's about one of your Georgia boys. You got time?"

"Sure, Chief, fire away. What else we gonna do?"

"Well, a guy named John Birch graduated from Lanier High in Macon where he lived. You ever hear of him?"

"No."

"Didn't think so. This is a true story. In '39, he graduated magna cum laude from Mercer. Then from a bible school, completing two years in one. In 1940, he sailed for China and mastered Mandarin Chinese in only six months. Now...if there's some absolute way to prove a person's brilliance, that's it. He was sent as a missionary seventy-five miles south of Shanghai which had been captured by the Japs three years earlier. The Japs kept moving southward, and Birch's bible mission was soon behind enemy lines."

"Why didn't he leave while he could?"

"He wanted to stay and teach the Chinese about Jesus; that was his mission...So, in April of '42, Jimmy Doolittle's B-25 crashed in China after his Tokyo raid. Birch was disguised as a Chinese farmer and was eating a bowl of dirty rice like a peasant at a street stall; that's how he blended in and survived. A Chinese patriot walked by and whispered that an injured American on the riverbank needed help. A patriot was someone who was against the Communists. Doolittle and his crew bailed out when their plane ran out of fuel, and Doolittle hid in a boat. Birch himself led Doolittle and his men through Jap-occupied jungles and villages to safety. Later, Doolittle told Claire Chennault, the leader of the P-40 Flying Tigers, what a tremendous help Birch had been. Needing a Chinese-speaking American who knew the country, Chennault made him a Lieutenant in the OSS; you know, the early CIA. Birch built a network of Chinese who informed Chennault with Jap movements and shipping. He was made a Captain and received the Legion of Merit and the Distinguished Service Medal."

"Just a typical Dawg," Pete said.

"He went to Mercer, moron," I countered.

"So, in August of '45, the Japs surrendered but were told to keep occupying their land until the Nationalists could take over. The Chinese Nationalists, not the Communists; we didn't want the Communists to take control when the Japs left. They're a ruthless bunch of bastards, just as bad as Hitler. Birch led some men to rescue a thousand prisoners still being held in a Japanese POW camp. Stopped by a squad of Communists armed with machine guns, Birch was told to give up his pistol. He said he was an American officer and the war was over, and his .45 was staying with him. He began yelling at the Communists, and his group tried to calm him down. Finally, he and a Korean were taken into the woods and shot. He had been shot in the stomach and was on the ground, and several Communists stabbed him with their bayonets. Birch's stomach and chest were covered with punctures, and his lower jaw was separated ear to ear from the rest of his face. The Korean had been knocked unconscious by a shot to his head, but he lived to tell the story. A secret investigation found that Birch's death was caused by the bayonet stabs while he was still alive, but our government didn't want anybody to know."

"Why not?" Pete asked.

"Well, listen. For over five years, Birch's mother was told he was killed by a stray bullet. When the truth came out, a Senator Knowland raised hell at the cover-up. He said Birch's death, if told when and as it happened, would have prevented the Korean War because our soldiers were already in China, and the Chinks wouldn't have had the balls to start the invasion. Some stinking Communists inside our government covered it up. Another Senator, Joe McCarthy, kept saying he had a list of over two hundred and eighty Communists in our government, but he could never prove it, so they got tired of him and ran his ass out. You guys never heard of Joe McCarthy either?...didn't think so."

"Did Birch start the John Birch Society?" I asked.

"No, but their only purpose is to fight Communism in honor of

Birch. Some people say they're a bunch of crazy fanatics who hate blacks and Jews, but that's not true. They only hate Communists who threaten our country. Birch was a good Christian who became fanatical himself, like a zealot. His men say he began holding his arms out like Jesus, and some say he purposely started the fight with his killers."

"Why'd he do that? He was outnumbered, and they had machine guns. Was he crazy?"

"Like I said, some people said he wanted to be a martyr to show the world how sinister the Communists were becoming. I like him; I like the man and what we did."

"Me, too," I said. "Chief, we know you're tired of us. See you later."

There was still a little light coming over the mountains on the edge of the harbor. Pete and I were worn out by the stories the Chief shared with us, but to our great pleasure. We walked to the flight deck and looked out at that magnificent city. Pete said, "Those Communist Chinese have control of almost all of this now. They will eventually force the British and other foreigners out. You wait and see."

"How do you know that?" I asked.

"I read a paper on the john every now and then," Pete laughed. "You dumbass. Hey, speaking of papers reminds me; did you see the one from Hong Kong somebody left in the shop yesterday? It said a company of crazed GIs shot up over 200 civilians in some village in South Vietnam. Man, the shit's gonna hit the fan!"

13. PEANUTS AND TRAIN STATIONS

The view from the ferryboats plying the dark blue water into Hong Kong was wasted on only the dullest of sailors. Motorboats and Chinese junks dashed around, busy in their teeming commerce. The high mountain tops surrounding the city were lush and green. The newly built hotels and shops glistened in the early sunlight. Hong Kong became an emerging center of corporate activity once the Communists permitted it. Businesses from all over the world clamored to build there.

Every type of merchandise from around the world was for sale in Hong Kong, but sailors received substantial discounts on top-quality goods at the Navy's enormous Fleet Exchange. After much thought about which to buy, I purchased a lovely diamond engagement ring for Jane, assuming she would have me. Both of us were ready, and a week later, the day after we left Hong Kong, a copy of our wedding announcement trimmed from my local newspaper was included in a letter from her. I was surprised to see the announcement, because I had not officially asked her to marry me, but she knew I would eventually get around to it. I was also surprised to see it was from my local newspaper, not hers.

The only problem with her engagement ring was I didn't have a place to hold it. Our Navy dress uniforms didn't have any pockets, front or back, only a slit inside the top of the pants forming a small change purse like that flap inside a bathing suit. I was afraid I'd lose the ring, so I put it on the little finger of my left hand where it stayed until I put it in our shop safe the next morning. The lack of proper pockets was only an occasional problem for sailors; we didn't carry a credit card, a driver's license, an insurance card, or a Medicare card. We didn't need 'em, and cell phones hadn't been invented. We only needed our military ID card and enough money to walk around, have a couple of beers and eat something different. We carried our cigarettes in our blouse, or shirt, pocket.

A cheap junket for tourists in Hong Kong was riding the tram up Victoria Peak to the top and looking around the harbor; that view inspired every visitor. One night just before Christmas, after consuming about a barrel of beer, Pete and I decided to go to the peak. A friend said eating a bowl of peanut soup from the man on the mountaintop was a treat. The old Chinaman had a big tub of soup boiling on an open wood fire. The pot sat on several large rocks in the middle of chunks of wood and ashes in a setting like a camp site. There were no tables, no chairs, and no utensils. But there was a scraggly line about ten people deep, all regular Chinese natives in shorts and flip-flops. For sailors on liberty, the man's famous peanut soup was the cure-all for the next day's hangover from a night of heavy drinking. For the Chinese, it was a complete, cheap, nourishing meal. The soup was nothing but heavily spiced cooked peanuts slightly thickened into a broth; we drank it, there was nothing to eat. Customers paid a nickel for the soup presented to us in a cheap porcelain bowl the man had just rinsed. He had rinsed it, not washed it, after his previous customers had finished their soup. But it really didn't even qualify as being rinsed; he just slid it quickly through a fifty-five gallon tub of brown water.

"Jeez. Did you see that tub? I wonder how long it's been since he's cleaned that thing," Pete said.

So I said, "How Long is a Chinese name."

"What do you mean? How long is a Chinese name? Five letters, ten letters? They don't have letters, you dummy, just a little chicken-scratching for each word. I don't get it; they're all the same length."

"It's a joke, Pete. Forget it. You said 'how long,' so I said 'How Long' with capitals. A Chinese man was named How Long. Boy, you're stupid."

I gave the old guy my nickel, and he filled my semi-clean cereal bowl using an old brown ladle with his right hand. With his thumb and the fingers of his left hand, inside and outside along the bowl's rim, he held it over the steaming pot, in case some spilled. He filled it flawlessly and handed it to me with a sweet but dingy smile. I said 'domo arigado' and moved to the edge of the mountain. As I stood beside Pete in the crisp mountain air, I slowly drank the peanut soup and hoped the earlier users of my bowl were free of cholera, diphtheria, diarrhea, shingles, chicken pox, measles, mumps, and typhoid fever. The peanut soup was a hit, two bowls was just right, and the tiny shirtless old man who apparently had cornered his market was to be congratulated.

Everybody on the peak went there for the soup or the view; nothing else was there. There was no security, and there was only one light, far-off and high up on a telephone pole. That didn't seem to be a problem; everybody up there seemed respectable. Hong Kong did have its beggars and homeless people, but just like in Japan they were neat, clean, and polite. They just needed a place on a sidewalk or a doorway to sleep during the night. I don't know what they did during the day. I hoped they enjoyed the Christmas holidays more than we did.

I received my 1969 W-2 from the Navy. My earnings had increased, thanks to combat pay, to $2,881 for the year, but that didn't include my free navy chow, my rack and fart sack, and the navy's available medical care. Everything else, my uniforms and college loan, I still paid from my take-home pay. I smoked and drank up the rest; I didn't save a dime for my upcoming wedding.

We returned on line on Yankee Station. On March 28, 1970 one of our planes, an F4J from VF-142 piloted by Lt. Jerome Beaulieu, shot down a MiG-21 North Vietnamese fighter. For the ship's crew and our aircrewmen, this was a happy day, something to break the monotony of the constant twelve-hour-a-day grind and the endless trips up and down the ship, and the same old food and the same old work. But for our flyers, our aviators, our officers, this daily grind was a deadly dance with death. Those brave American men put their lives at risk every time they entered their cockpits. Our squadron pilots and bombardier/navigators had done their duties admirably and had fared well, but our cruise was not to end until May.

In mid-April we pulled into the harbor at the U. S. Navy base in Yokosuka, Japan. Sprawled out on the Mura peninsula in Central Honshu, Yokosuka was the largest overseas U. S. Naval institution in the world and was considered to be the most strategically important base in the U. S. military. When our ships pulled into any Japanese ports, the smell of ozone from electrical torches cutting through steel during ship repairs was harsh and bitter; a sharp smell like progress, like money, like industry. In the winter months, the Japanese harbors were usually crowded, overcast, gray, damp and noisy. They were loud, and gloomy, and magnificent.

Back in Japan again, we sailors were urged to remember that no other place was like Olongapo where almost anything was acceptable. In any port of call besides Po City, we tried not to drink more beer than we could handle so we would act semi-properly. In ports other than Olongapo, we concentrated on culture and exotic foods and unusual goods; purchasing tape decks, stereos, and cars which cost a fortune back home. They were cheaper here; there were minimal taxes and shipping was free, thanks to the good ole Navy. Pete and I decided to spend a few dollars travelling around a couple of old Japanese cities by train and finish up in Tokyo. Early the next morning, we hopped the express train to Yokohama.

Neither an Officer nor a Gentleman

The strangest thing I saw during the entire four years of my naval career appeared like a mirage before me. We stepped from the train and followed the busy morning crowd walking shoulder to shoulder into the Yokohama station. A plain, flat platform sat in the middle of an enormous lobby the size of a college basketball arena, but the station was seedy and gray and had no works of art to impress visitors.

There were at least forty Japanese men standing near the stage. The top of the platform was at Japanese eye-level. Pete and I didn't notice what was going on even though we were a foot taller than everybody else. We had no clue what was happening as we worked our way through the crowd toward the doors that led out of the station. The men parted a little, and we unintentionally walked right up to the platform. On its front edge, ten feet from us, a completely naked Japanese woman was lying on her side facing the men. She propped her head up with her arm, her jet-black hair cut short in a bob. She had a lovely smile; her white teeth were perfect. Her slim, naked, cream-colored body was resting on the bare platform with a sky-blue baby blanket under her. She must have been a little cool; her small Japanese boobs stuck out like little pink Hershey kisses. She wouldn't have to worry about them getting flabby in her old age; they would probably just disappear. It took a minute for Pete and me to take in all we saw. It was surreal; we were standing in the middle of a public train station.

The girl looked at Pete with her beautiful almond-shaped green eyes and rubbed the smooth skin of her leg. She said something in Japanese and motioned with her finger for Pete to come near. The Japanese men all got excited and started chattering with each other. Some of them laughed and turned to look at Pete. We just stood there because we had no idea what she said. She repeated it in that high-pitched sing-song way the Japanese women talked. Amazingly, I thought I caught the gist of it the second time. The men around us turned and pushed Pete toward the edge of the stage. The girl smiled and rubbed her leg again.

Pete inched up to the woman and just stood there staring at her lovely body. She said something again to him in broken English, and I got it that time.

I was in a trance. I felt like we were the butt of some crazy joke, but I couldn't figure it out or do anything about it. I just wanted to get out of there. I had never seen anything like that before, and I was a little scared and embarrassed. Pete was playing this strange game in a public place, and I had no idea what would happen. It was ridiculous that I, a Navy man whose sole existence at that time was to enhance the killing capability of U. S. military jets, was embarrassed by such a petty thing. I felt like a fool standing there in the middle of that crowd, looking at that naked woman in that train station in broad daylight. I had no earthly idea why she was there.

I had seen enough and yelled, "Pete, let's go!"

He glared at me, then turned and stared at her one long, last time, committing her wondrous image to memory, and slowly began to leave.

As we separated from the crowd still glued to the stage, I asked Pete, "Why didn't you do it?"

"Do what?" he said.

"You heard what she said."

"Yeah. I heard her, but I couldn't understand a single word. What did she say?"

I looked over and laughed at him, "You really don't know? She was telling you to kiss her. That little high-pitched thing she kept trying to say was, 'Kiss me, sailor.'"

Pete said something nasty and turned around. "Let's go back."

I said, "No, let it go. You had your chance, Romeo. It's gone; it wouldn't be the same."

We looked around the city and drank a few beers. The train to Yokohama was fast but nothing like the bullet train we took to Tokyo. We spent most of our time in Tokyo in the Ginza district. Sexpo70, part of Osaka World's Fair, was the big international attraction at that time, and Tokyo had major spinoffs. The skits were almost soft porn, but I had been away from prim and proper company for so long that I wasn't a good judge. I knew many church members who wouldn't have approved of a single one. The Japanese have an unusual sense of humor, so most of the shows were centered on sketches of characters like Snow White, Little Red Riding Hood, and Sleeping Beauty dressed in beautiful, colorful, flimsy, and sexy outfits. The poor girls were usually involved in some humorous but compromising situations. The grown-up every-day Japanese seemed to find great humor in the silliest of things, almost like children. They roared with delight at the skits. Osaka was a little too far and expensive for a day trip, so Pete and I didn't go.

Sometimes, we got a little tipsy and acted like children ourselves. We drank just to have something to do. One night we ran into some young Japanese sailors about our own age and started hanging out with them. Exchanging uniform hats was a bad idea; we weren't kids, we were professional military men. We shared cigarettes and laughed with each other just like old friends. No one understood a word that the other guy was saying, so we communicated by making faces and gesturing. Whenever that didn't work, we pulled out a little pocket English-to-Japanese dictionary Pete carried. The Shore Patrol saw us acting like the fools we were and broke up our little party. We had to return their hats which made me mad; I wanted to keep one. They looked like the little round black hats with ribbons on the back that Donald Duck and his nephews wore in the cartoons.

Often, while we drank in some bar, some Japanese World War II veterans about my daddy's age approached us and would want to show us how strong and defiant they still were. It was all in good fun on our part, but they took it a lot more seriously than we did and would have physically kicked our butts if given the chance. They hated they had lost the war, it had only been twenty-five years earlier, so they acted like some old flag-waving Southerners who were still pissed off about the Civil War. They'd stare at us for a while, and after a couple of drinks, one of them would roll up his sleeve and gesture to us, and we'd end up arm wrestling. We'd take on all challengers until they eventually got mad. Someone in the bar would break it up before they became belligerent. Pete and I were larger than they were and had been lifting heavy airplane parts every day for a year, but they wanted to show us they were still full of fight. They were full-grown longshoremen who lifted heavy sheet metal and ship parts all day, so sometimes we lost.

Most of the time we did crazy stuff just for fun and didn't cause any trouble, but occasionally we did cause trouble. Several times Pete and I played with the restaurant staff; it seemed funny at the time. Nobody tipped a waitress in Japan; it was not their custom. We found that out when we innocently left our first tip and our waitress handed it back. We explained in hand gestures that the money was for her. She just shook her head and refused to take it, so we thought it would be fun to start hiding our tips under our plates whenever we left a restaurant. We soon saw the trouble we were causing. Instead of just pocketing the money or turning it in to the cashier, one little Japanese waitress came running after us as we walked down the crowded street. She didn't know why we had put the money under our plates, but she returned it to us anyway. We had the decency to quit wasting their time with that crap after that.

We walked back to the train station to return to Yokosuka and then the *Connie* after a long day. We discovered the trains had stopped running at two in the morning. We were cold, tired, hungry and disgusted. We had run out of money, but fortunately we had bought return tickets before we left. To keep warm, Pete and I began walking like we were marching in a parade. Up and down the platform, calling the cadence out loud. A policeman at the station came up to us and asked us what we were doing. I told him our story and said we were freezing. He said to follow him as he led us to the station master who graciously invited us in. His small office was nice and warm. He was a retired American Army sergeant who had been in Japan ever since they had surrendered to end World War II. He had been home once in the last twenty-five years. The fifty-year-old sergeant reminisced about his times in Japan and served Pete and me some hot green tea. We whiled away the next two hours talking about his many, many Japanese love conquests until the trains began their service again. I didn't appreciate his bragging about deflowering twenty-six inexperienced adolescent Japanese virgins like a giant frenzied shark slicing through a school of baby mullet. We thanked him and wished him well; he had been nice to let us share his warm office, so we bowed deeply and offered a formal 'domo ariga-do' as we left.

The middle of April of '70 found us in Subic Bay taking on fuel and food for the cruise back to Hawaii and San Diego. We were going to have one last fling in Po City. We were worn down after our eight months at sea, by our endless twelve-hour days without a break when we were on line, and by all the drinking and partying we had done. We were all tired of the war and the bombing and wanted to go home and see our families; we had all had enough. I wanted to see Jane and kiss her lips and smell her hair.

The once-upbeat gung-ho sailors coming off the ships on the way to Yankee Station were not the same on their return trip home. Olongapo was comparatively quiet during liberty call this last time. Some of the men didn't even go ashore. They did leave the ship, but only because we always had work to do whenever we were in port. Our planes flew in to NAS Cubi Point adjoining Subic Bay, and we worked on them just like they were on the flight line back at Oceana. It was almost like being home, except for the heavy drinking, the wild partying, and the wanton women. Okay, it was just like being home for some guys. Pete and I took one last trip into town and drank a couple of beers. No women, no fights, and no trouble.

One of our older shipmates had invited us to a farewell party at a house he usually rented whenever he was in Olongapo, and we stopped by to say goodbye to him and his Filipino lady friend. Some sailors rented a house for the duration of their stay in port, and some not only rented a house but also a woman, and each for almost nothing. That was really the embodiment of renting a house fully furnished. Their lady friends cooked for them and washed for them and made them happy in many ways. It was not an uncommon occurrence; to them it was like being at home with their wives.

My last good memory of the Philippines was set there in that rented house. Our older shipmate's girlfriend had prepared a special national Filipino dish for him and his buddies which just marginally included us, so we didn't stay for supper. It was her special way to say goodbye to him until either his next cruise or until he transferred to another West Pac squadron. This guy would come back again because he was a lifer, a career military man, and he would request West Pac duty until the war was over. We young squids needed our butts kicked for calling those older sailors 'lifers' but we were just plain stupid. Those big, healthy, middle-aged men outweighed us by fifty to a hundred pounds and could have physically beaten us to a pulp. We didn't have enough sense at that time to realize that they were honorably serving their country as they were called to do and what a sacrifice they were making for all of us.

Pete and I didn't stay long; we were his good, but not his best, friends and left just as the main course was brought from the kitchen. As I turned at the door for one last look and to wave goodbye, I saw his lady friend put a small American flag on the specialty dish sitting in the middle of the dining room table. She didn't put the flag on the platter that held the specialty; she put the flag on the specialty itself. She stuck that small American flag into the back end of the roast dog sitting in the middle of the table. The dog had a big, red, cooked apple in its mouth just like a suckling pig. She had thoughtfully cut off the dog's tail so the flag could proudly wave straight up in the air. She placed the tail which looked like a long chicken neck along the side of the platter. It was a fairly large dog; I always wondered what kind it was.

Our nine-month deployment to the western side of the Pacific Ocean with Carrier Air Group 14 ended on May 8, 1970 as we unloaded at the pier in San Diego. The air wing had lost seven aircraft, five to enemy fire, during the cruise, but our VA-85 squadron came home with all our planes. No aircrewmen on those seven downed planes died, but one aviator was not rescued when he bailed out of his flaming jet and became a Prisoner of War. Twenty American sailors and one civilian on board the *Connie* died during the cruise. At the back of the 1969-1970 *Constellation* cruise book, in a place of honor, the "In Memoriam" page lists their names and units: AM3 Roy G. Fowler (Medical), MM1 Paul E. Gore (P-1), ABH3 M. D. Gorsuch (U-2), FTM2 R. B. Leonard (Fox), ABF3 Brian R. Lighthart (U-2), AZ3 Scott F. Moore (HC-7), DS3 K. L. Terrell (S-7), Mr. F. L. Bytheway (Civ), AMH2 Carl I. Ellera (VA-97), SD2 Fidel G. Salazar (VAQ-133), ASE3 Michael Bowman (VA-27), HM2 Donald C. Dean (VAM-113), AME3 Terry L. Beck (VF-143), AE2 James J. Fowler (VF-143), AQB2 R. M. Montgomery (VF-143), ATR3 Richard W. Bell (VF-142), AMS3 D. L. Kohler (VF-142), AN H. M. Koslosky (VF-142), ADJ2 K. M. Prentice (VF-142), CDR Randall K. Billings and LTJG William L. Beaver. Our one pilot "Missing in Action" was LTJG Jim Bedinger.

The crew's photos were still shown in the cruise book for those men who died after they had their pictures taken. No sad words or negative pictures were in the cruise book because the sailors of the *Constellation* wanted to only remember the good times. There were no explanations given in the cruise book as why or how any of these men died, but they all had one thing in common: they were serving their country's call in the midst of a war when they died. They were all members of that honorable group of men called sailors, except for one, Frank Bytheway, a civilian working at NAS Cubi Point, PI. Frank LeRoy Bytheway was born in Utah in 1934 and was lost at sea October 2, 1969. He was an electrical engineer at Cubi working for Collins Radio, based in Manila, and was headed to the *Constellation* to work on the ship's radar and communications system.

By the way, along with the above-listed 17 members of the *Connie*'s crew, five members of the plane's crew and four other sailors reporting to other warships were lost at sea and their bodies were never found. The crew of the C2A Trader cargo aircraft from Reserve Cargo Squadron 50 assigned to transfer the men from Cubi to their ships were Lt. Herbert Dilger, pilot; Lt. Richard Livingston, co-pilot; AMS3 Rayford Hill, crewmember; ADJ3 Paul Moser, crewmember; and ADJ3 Michael Tye, crewmember. After takeoff from Cubi, Lt. Dilger reported "Ops Normal" which was confirmed by other squadron aircraft and the *Connie*'s air control center. Approximately 55 minutes later and 26 miles from the *Constellation*, radar contact was lost. The plane went down 68 miles due east of the North Vietnamese coastline in the waters of the Gulf of Tonkin, and an extensive SAR operation immediately began. A chopper recovered a few pieces of the plane, but no bodies of the crew or passengers were found. The recovered debris indicated that the plane went down in a relatively high-speed nose dive, right wing down at impact, or a possible right wing failure before impact.

Two of *Connie*'s men died on this cruise from making the fatal mistake of not being careful; one walked into a turning propeller and one was eviscerated by a jet engine. The other man was the young black sailor we buried at sea after being burned to death.

Another death, that of Commander Randall K. Billings, occurred in a tragic, non-combat related incident. Traditionally, commanders of

Navy air wings flew all the aircraft assigned to them, and Commander of the Air Group (CAG) Billings was only checked out on two. Billings, commander of CVW-14, asked to fly a Vigilante while the *Connie* was at Cubi Point for the holidays. LTJG William Beaver was assigned to take the flight with him.

For some unknown reason, the Vigilante went into a steep dive with its engines at high thrust. Beaver's radio transmissions showed he tried frantically to talk to CAG, and when there was no response, he ejected. In the RA5C there is no system where the Reconnaissance Attack Navigator (RAN) could eject the pilot, so Beaver could not eject CAG. The pilot could eject the RAN, but CAG was unresponsive. The jet was going faster than Mach 1, and Billy Beaver in the rear seat released his seat connection while trying to regain communications with CAG in the front seat. Beaver gave up and ejected himself, but the jet's high speed and his unfastened seat connection caused the ejection to break his neck, killing him. The jet made such a large hole at impact that CAG's body was never found.

The *Constellation* had a long, tragic history of brave aviators being shot down and captured during the war. The first American aviator, and probably the most well-known, to be captured by the North Vietnamese was LTJG Everett Alverez (VA-144) who launched from the *Connie* on August 5, 1964. LTJG Alverez's A4C Skyhawk was shot down just three days after the Gulf of Tonkin (also known as the U.S.S. *Maddox*) incident and endured eight years and seven months as a prisoner of war in the Hanoi Hilton. A year before Alverez's release in February of '73, two of Connie's pilots, LT Randall H. Cunningham and LTJG William Driscoll, became the first American aces of the war by shooting down five Russian-made MiG jets. Some of those five were downed on May 10, 1972, when the *Constellation*'s VF-96 air crews shot down six MiGs, three by Cunningham and Driscoll in their F4J Phantom.

F. Lewis Smith

The Prisoner of War noted on the "In Memoriam" page of our cruise book was LTJG Henry J. Bedinger (VF-143). On November 22, 1969, LTJG Bedinger and his pilot, LT Herbert Wheeler, were launched from the *Connie* flying their F4J Phantom fighter. On their reconnaissance mission over Laos, their aircraft was hit by anti-aircraft fire as they were over Savannakhet Province 10 miles from Sepone, and they were forced to eject. Both parachutes were sighted and contact was made with both on their survival radios. LT Wheeler was picked up by SAR helicopters, but heavy ground fire prohibited LTJG Bedinger's rescue. His last radio transmission said, "I guess I'm a prisoner of war." Oriental chatter was then heard. Bedinger had been captured by the NVA operating in Laos and was immediately moved to North Vietnam where he spent the next three and a half years as a POW. Bedinger was separated from other Americans who had been captured in North Vietnam. At the end of the war, Bedinger and another dozen Americans were presented as "Laos prisoners." Bedinger had spent only 10 days in Laos, primarily the time required to travel to North Vietnam. After much Governmental haggling, the men who were captured in Laos and were immediately taken north were added to the list of returnees being released by the North Vietnamese. Jim Bedinger was finally freed and in the '80s returned to work in the office of the Joint Chiefs of Staff at the Pentagon.

14. WEDDING BELLS AND WIVES

Removing our VA-85 gear from the *Constellation* took a day and a half. Pete and I needed to find a way home to Georgia because the Navy only had free passage for us on planes to Virginia Beach, and they left in five days. We cleaned up after our last sea chest was stowed on a cargo trailer and made our final departure from the ship. Six of us stood on the pier waiting for a cattle car heading to San Diego, and one of our officers walked up. "Any of you men going to Texas?" he asked. "I've got a Navy plane with four seats available if you can use one. It won't cost you anything."

I looked at Pete and said, "Yes, sir, we can use two of them."

"You guys from Texas?" he asked.

"No, sir," I said, "We're from Georgia, but that'll put us half way there."

"Great. Put your seabags in that black van over there. I'll be there in a minute."

When the van stopped at Halsey Field on NAS North Island, we climbed aboard an old Douglas DC-3 and took a seat. The seven hour flight got us into Waco an hour after midnight Texas time. We grabbed our bags, thanked both the pilot and our squadron officer, and started for the hangar when our officer yelled, "Hey, men, hold on. Let's see what I can find for you."

In two more hours another DC-3 was flying to Robins Air Force Base nine miles from Macon, a four and a half hour flight, and we were on it. We landed around six-thirty the next morning on Georgia time. We took a cab to the Greyhound bus station and bought two tickets to Augusta. The next bus to Augusta left at eight-thirty so I called my daddy and asked him to pick me up, and Pete called his mother. As we stood chatting on the corner of Greene and 12th Street, waiting for our rides, I noticed a large dark column of smoke coming from downtown Augusta.

"Man, it looks like some building is burning," I said.

Midge picked Pete up, and I said I'd talk to him soon. Fifteen minutes later, my daddy hugged me hello and apologized for being a little late, "Sorry, son, I had to take a detour. The blacks have taken over the city and are tearing it up."

"What did you say?" I asked as I slid onto the front seat. My left hand touched something hard and cold as I got settled; my daddy had his Army Air Corp Colt .45 on the front seat, right beside him. "What's this for?"

"I'll show you," he said.

We pulled out of the station and headed toward the fire. Two blocks closer, we saw the first evidence of a riot. Houses and buildings were looted, cars were overturned, and trash was everywhere.

"You're not driving through that, are you? I just got out of a war zone and don't wanna see another one. Let's go home; I'm tired."

Daddy ignored me and turned onto Broad Street near the main business district. I looked down a side street and saw a hundred black people standing in front of the police station. Daddy said, "Hell, that's nothing; there were twice that many yesterday."

The black citizens were trying to find out how an inmate had died. Daddy said, "Everybody was told he was killed by the police; it was in the *Chronicle* and on the TV, but he was really beaten to death by some other inmates over a damn card game."

"He was beaten to death by some inmates? Over a card game?"

"Yeah. They were playing some card game I guess they play in jail. They don't have any money, so they play that whoever loses is gonna get his ass beat. A retarded boy named Charles Oatman lost and was kicked and beaten by the other guys. The poor boy climbed unto his bunk to escape, but according to the police, he fell and crushed the back of his skull. Later, two black inmates, two grown men, confessed to beating him. Can you believe that crap?"

My daddy had picked up his bad language in the Army Air Corp during World War II, and I learned mine from him as a youngster. I had gotten my master's degree in it as a sailor during Vietnam.

"What was a retarded boy doing in a cell with grown men? Why wasn't he in a youth detention center?" I asked.

"Exactly," daddy said, "that's what everybody wants to know."

The men confessed; the rumors that the boy was killed by the police were false, but the damage was done. People were saying that the boy had been beaten often and severely by the other inmates, and the problem was the police knew it and did nothing to stop it. That was the true injustice of it all. Activists and preachers fanned the flames, claiming injustice and police brutality, until a riot broke out. A white man was pulled from his vehicle on Ninth Street and was beaten until two black men pulled the rioters off him, saving his life.

Police Chief Broadus Bequest initially decided the police were outnumbered and didn't try to control the rioters, but eventually he turned his men loose. Governor Lester Maddox had offered the city the services of the Georgia National Guard at the very beginning, but city officials said they could handle it. At two a.m. on the twelfth of May, the Guard arrived with their tanks, but the rioting had just about played out.

"I read that the Guard unit stationed beside the Augusta College campus was called up. You know some of those boys, don't you?" daddy asked.

"Sure, Jack Fisher and David Scott and Randall Davis .... and Robert Bullard, everybody who thought they might go to Nam joined the Guard as soon as they could; I probably know twenty of the boys. They thought that would keep them from being drafted into the Army. Did you read that they were the first unit called up for active service? How's that for a kick in the butt? I think they just got back, and now you say they're involved in this? Man, I feel sorry for them, but at least they're experienced. They'll know what to do."

Surprisingly, my daddy, never known as being a big advocate of civil rights, said, "In a way, this was a good thing. People always say Augusta doesn't have a race problem, but now they know we do, and something's gotta be done. You remember James Brown, the singer, whatta they call him, the Godfather of Soul; he came back to town from a concert just to help calm everybody down. I think he's a good man. He's not Hank Williams or Merle Haggard, but I think the blacks and the hippies love his music. I'll take you by his house when we go home. It's really a big mansion over there on Walton Way by the Augusta College president's house. I bet the neighbors are really pissed off. I heard they offered to double his money if he would sell it and leave."

"Hippies don't love his music, daddy; they hate it. They like that weird Jefferson Airplane psychedelic stuff like we do."

"We?" daddy said.

"Yeah, me and my buddies. We like the Godfather and other black music, but we like the Beatles and the Stones and the Four Tops, too. Stuff like that. You like Perry Como and Frank Sinatra and crap like that. You'd like Harry Belafonte, too, but you heard he's black and dates white women."

Before the shooting stopped, six black men were dead, all shot in the back, sixty-two were injured, and more than twenty buildings downtown were burned.

After a few days in Augusta, Pete and I returned to Virginia Beach to help install the newest avionics systems in the fleet in our jets. We returned to chill out and enjoy the sun and the beach and to renew friendships. It was a time for hotdogs and hamburgers, steaks and eggs, movies and dances, beer and wine, love on the moon-lit beach, and playing golf with friends. There'd be no wild stories for a while, just delicious Lynnhaven oysters, little bitty Western Sizzlin' and Bonanza steaks, and trips to DC. And always fun, sun, music, beer, and wine. Guys left our squadron; some were sent to new schools and some retired. We trained their replacements and prepared for another deployment.

My momma wrote me a letter a year earlier when I told her I was marrying a girl from Pittsburgh. She knew my wife was a fine woman, she said, but she worried we would move up there, and she wouldn't get to see her grandchildren. My daddy's brother had done the same thing to my grandmother, so I knew where she was coming from. I told her not to worry, and she never mentioned it again.

Jane taught at an elementary school and lived at home with her step-mom. Jane's birth mother had died when Jane was five, and her only brother was killed in a head-on car wreck on top of a bridge one cold night when Jane was eleven. Her father died from a massive heart attack during a business conference in Virginia when she was fourteen. After their father died, Jane and her sister Elizabeth lived happily and well with their step-mom and her four kids. Not only was Jane blessed with brains, her nickname in high school was 'the Body.' The only complaint she had in life until she met me was her step-mom wouldn't let her shave her legs until she was in the eleventh grade.

Jane was a brilliant student and graduated with honors from her tiny Dunbarton College of the Holy Cross in DC. Her roommates and friends at school were all very bright and personable; they all achieved much success in their later lives. She was apparently the only one of the group who was not rich, much to my surprise. Her friends were proud of her like she was of them. The girls graduated and scattered all over the country. Jane was going to miss them in time, but right now she and I found so much pleasure in each other's arms that we didn't need anything else.

After our marriage, my momma became the real mother Jane missed in her life; they loved each other until momma died. Every time we all got together, Jane was by her side cooking and cleaning and always laughing. Jane told momma after a disappointing visit to Pittsburgh that no one seemed to care if we were there or not, and she wondered if she'd go back. Momma said, "Well, you can just stay down here with us. We'll be your family."

Momma always treated Jane as if she were her own flesh and blood, and my siblings all accepted her as a major part of our family, more than just an in-law. Jane always felt she totally belonged with us and usually organized our family get-togethers.

In June of '70, Jane and I were married in Sacred Heart Catholic Church in Emsworth, PA. The priest who married us was a family friend only five years older than Jane. Our reception was at the University Club at Pitt; Jane's father had been a member. We left the reception at four o'clock and headed for two weeks of sightseeing in Maine and Canada; then to DC and back to Virginia Beach. We stayed in a motel near Oceana for ten days because I had foolishly not told base housing in advance that we wanted to live on base. That was one of the many stupid things I have done in my life.

Base housing had everything a young couple needed. We spent fifty dollars on cheap pictures, rugs, towels, knick-knacks and stocking the refrigerator. Downstairs, the apartment had a decent sized living room with a sofa and matching chair, a tiny dining room and a modern kitchen. Upstairs had a nice bedroom with a double bed and a small bathroom with a shower. The bedroom faced the ocean and had a large picture window with a smaller screened window on each side. The light curtains covering the windows billowed sideways into the room inviting the cool morning breezes to flow all around us. We lingered in bed every chance we got, drinking coffee and nibbling on Jane's cinnamon rolls. On the weekends before we started our day, we would read the morning paper in bed. We were in love and felt at home.

Sightseeing in Richmond and DC and Williamsburg, going to concerts featuring Dionne Warwick, the Temptations, the Four Tops and the Supremes in Hampton Roads and Norfolk, swimming and drinking beer at Dam Neck and Little Creek while listening to Jose Feliciano and the Doors on the radio, we did it all. We went home to Augusta to see my family and Pittsburgh to see hers where we ate bratwurst and Braunschweiger, kielbasa and golumpki. My brother Jimmy and his wife Sue drove to Augusta from Alabama to visit us. While drinking beer in the back yard, she asked what I thought about the Lt. Calley case. I said I hadn't heard about it. She said, "Oh, I thought you knew. You know poppa works at Ft. Benning and all the officers are talking about it big-time."

"Never heard of him; who is he?"

"Well, poppa says he's an Army lieutenant who led his platoon into some village in Vietnam and started killing all the women and children. In cold blood. Unarmed civilians."

"Viet Cong?" I asked.

"No, poppa says they were just old men, women and children."

"Jesus," I said, "this shitty war is ruining us all."

We returned to Oceana feeling young and happy. We were living on a shoestring but were much in love. What a wonderful creature Jane was. Pete came by often, and we were glad to see him. Jane liked him a lot, and they got along well. It was hard to believe our Mediterranean cruise was taking shape on the near horizon. It didn't matter to me which ship I was going to be on. Sailors who made the Navy a career got to know them all and had favorites, but we short-timers didn't.

Sailors called the U.S.S. *Saratoga* (CV-60) the "*Sorry Sara*" because she wasn't as spit-and-polished as the newer carriers coming into the fleet. Rumors said she was our next ship, but my squadron was ordered to the U.S.S. *Forrestal* (CVA-59). She was commissioned in '55 which made her only sixteen, and we heard she was in great shape. We were happy with her, which was good, because we had no choice.

The *Forrestal* had been deployed on nothing but Mediterranean cruises since she was launched, but finally made an appearance in June of '67 on Yankee Station in the Gulf of Tonkin. According to Navy history, she was in the fifth day of her first combat cruise when disaster struck, killing 134 men. It was the same old story. A jet turned, and its exhaust flames ignited a nearby jet's missiles. Those missiles flew across the deck, slamming into more jets fully loaded with missiles and bombs. Exploding rockets and burning jet fuel was everywhere. The flight deck was awash with flaming fuel, bleeding down into the decks below, killing sailors asleep in their bunks. Lieutenant John McCain scrambled out of the cockpit of his jet which was ready and waiting for launch and calmly walked out unto its nose and jumped onto the flight deck. He escaped with no injuries during the disaster, but would suffer mercilessly later, at the hands of the Communists at the Hanoi Hilton prison camp in North Vietnam. Damage was so bad from the fire that the *Forrestal* was sent back to Norfolk to be repaired.

The heavily repaired *Forrestal* was deployed to the Mediterranean in July of '68 and returned to Norfolk in April of '69. She left again in December and returned in July of '70. As customary, she was cleaned and painted fore and aft by elbow-grease and battleship gray paint. All routine and necessary repairs were made, and she was ready for sea again.

The *Forrestal* held her Family Day Cruise in October of '70, and over five thousand family members and friends joined us in a day-trip way out into the Atlantic, giving the Navy dependents a chance to see what life aboard a carrier was like. It was my pleasure for Jane to join me; I loved showing her off to all my friends. She wore her long, thick hair pulled back almost like a ponytail and had a silk ribbon wrapped around her neck. The most striking thing about her were those long, beautiful, still-tanned legs busting out of her little mini-skirt. Her figure was so fine that it was exciting to see her in a bikini. I had several men, total strangers, ask me who she was. When I told them she was spoken for, they just nodded their heads, giving me their silent seal of approval. One of our aviators looked sideways at me and said, "You're one of my boys, aren't you?"

Pete hadn't invited anyone to join him on the cruise and left us alone after a brief hello. The weather was upbeat, warm and pleasant, and so were the spirits of our visitors who forgot for that day the long period of isolation staring them in the face. Navy wives and children, especially at that time before cell phones and the internet, had a hard time when their men were at sea. A military life was fine when the family was all together, but when one spouse was not home for their children's graduations, birthdays, or confirmations, it was tough on both spouses. Fifty years ago, that 'one spouse' meant only men; there were no women at sea, but that's all changed now. Sickness, death, and legal problems were hard situations for one spouse to handle alone. Discipline problems and drug and alcohol abuse were common occurrences in families where a parent was missing, military related or not; the same as today. It was hard to reconcile being at sea for nine months and then being home for nine months and then leaving again for seven or eight months. That process repeated for twenty years would generate many serious problems, not all dealing with fidelity.

A few VA-85 personnel had their wives follow them to different ports of call when we were deployed. I presume it took a lot of money to afford that; maybe combat pay could swing it, but there would be no combat pay on this upcoming Mediterranean deployment. Aviators made more money than we peons; maybe they could afford a travelling spouse. They also received flight pay for sticking their necks out every day and certainly deserved every penny. For the rest of the guys, it couldn't be done without a rich momma or daddy or spouse to bridge the money gap. Two enlisted men in our squadron had their wives follow them, thanks to their inheritances. It was a dangerous journey for those wives who did attempt it. They had many terrible experiences dealing with foreign cultures and languages and people they encountered along the way. Many times in other countries, women were mistreated because they were not considered equals to men. In some countries, they were considered property and had no human rights at all.

A wife following her husband from port to port had to find a place to live while her man was on duty at sea. But unlike ordinary tourists, a Navy wife was occasionally subjected to sudden changes in itinerary due to rescheduling of cruises caused by weather or world events. In some circumstances, a traveling wife had to deal with the injury or death of her loved one. Wives were sometimes told their husbands had been shot down and were missing in action or were dead. Those things were horrible to handle even when wives had support from family members, but being in a foreign country with just another Navy wife or a chaplain to comfort them was catastrophic. As helpful and loving as her girlfriends and other Navy personnel were to a grieving widow, they were not a substitute for family.

Everyone knew to be happy on the family day cruise, no matter how hard it was. The PA system boomed out, telling us to hurry to the flight deck to see the annual air show presented by our flyboys. A great procession of about twenty-five of the *Forrestal* jets in all makes and sizes zoomed overhead, some low and slow for the benefit of the kids. The best part of the show for me was the Vigilante's flyby; it broke the sound barrier right above us on the flight deck. None of the dependents were prepared for that, and neither were the rest of us; half the people jumped a foot high and squealed with delight or horror. Some of the littlest kids cried uncontrollably. The cruise was a nice gesture on the Navy's part, which we sincerely appreciated. It let our family and friends see where we would soon be living and working, albeit unhappily without them.

15. THE FUNERAL GUEST

The Wednesday morning after the family cruise, Pete got a six o'clock call from his mother's neighbor in Augusta telling him the poor thing had died three days earlier. The neighbor had to go through Family Services to locate him. Unfortunately, we had planned to go to a concert in Richmond that weekend, so any upcoming funeral was going to ruin his plans. There was no question he had to go, he was an only child, so he called me to say he wanted me and Jane to go along to comfort him. Comfort him; what a joke, he needed no comfort, there wasn't an uncomfortable or sad bone in his body. The only thing he wanted from me was to drive him there and pay for the gas. What a helpless pain in the rear he was, but we were certainly going.

Before the Chief could get too busy in the shop, we went straight to see him to get our liberty passes. As thoughtful as ever, he told Pete how sorry he was and said to call him personally if there was anything he needed. He looked at me and said, "Be careful on the road, hot shot."

"Thanks, Chief; we'll be back on time."

"Don't make me come after you half-wits again. I'm too busy this week to waste my time down in the swamps looking for your butts… in those damn alligator-infested jungles and hundred and fifty degree heat and mosquitos big as horseflies."

"We're Georgia crackers, Chief, not Floridians."

"Oh, yeah, there's a big difference," he smirked. "Get out before I change my mind."

Before we left, we sold our tickets to the Jimi Hendrix concert to a shipmate in the shop at half price. Chief Cranston let Pete call information and then Chicago to his uncle Toogie from the squadron office. When told his sister was dead, Toogie wasn't too upset; he hadn't seen her in over four years. He told Pete to relax; he would talk to his childhood buddy in Augusta, a big attorney and politician now, and handle everything. We said a prayer of thanks and left the base before ten.

Pete's mother had been quite popular when she lived in Country Club Hills and had a husband and some money. She had hosted dinner parties and evening socials. She had filled cups to overflowing with the best wines and champagne. Her homemade, delicate desserts were better than the best that money could buy. She was obsessed with proving she was somebody special. When her husband died, she quickly ran through his money still trying to prove it. She yearned for love and acceptance and recognition from other humans, even if it were just a nod of the head or a simple thank-you. She had never received any of those from her son. In a dreaded financial bind, she sold her fashionable house and moved to an older two-bedroom place three miles away.

Her friends and neighbors didn't forget her when it came time to lay her to rest. As she became older and less dependent on material things, she had become a regular at the Methodist church down the street from her Bluebird Road home. She didn't join the church because she had experienced a life-shattering epiphany; she just wanted some plain old female companionship. Nothing had happened to make religion the center of her life; she wasn't too worried about saving her eternal soul. The many church members at her funeral would have made her happy. She had always paid her church 'dues,' she once told Pete.

It was a beautiful fall afternoon on Saturday when the funeral procession convened at Westover Cemetery. Jane and I were there, but we had not seen or talked to Pete since arriving in Augusta. We had spent the past three nights at my mother's and had flea-marketed in Atlanta one day and Columbia the next. Pete had rented a car and was staying at his mother's once happy home. The preacher from the church was there, as were a few family friends. Midge's brother Toogie and his wife sat in front next to the seat left for Pete. Six rows of people waited patiently under the large green canopy adorned with 'Overton Funeral Home' in white letters along its sides. The service was ready to begin, but it seemed that we were waiting for Peter Boy.

Pete did not have any funeral clothes at Oceana or in Augusta, or anything else that was appropriate to wear to his mother's funeral. He had not even considered looking for some, or buying some, or asking to borrow some from me. On the morning of the funeral, as he finally rummaged through the house looking for some of his old clothes, he decided to drink a few beers to calm down and smoke a joint to clear his head. On a shelf in the guest bedroom closet, inside a wedding-dress box labeled 'Pete,' he found his old Senior Prom outfit that his mother had saved. He was feeling exceptionally mellow at that point and put it on. It fit almost perfectly, and he thought he liked it, whether he was high or not.

When Pete sauntered down the aisle to his seat under the canopy, he was fifteen minutes late and a little unsteady. He had come to his mother's funeral wearing his new-found wedding box inheritance; the white suit, white vest, white shirt, white bowtie and white shoes she had saved for him. As accessories, he brought several other items he found around the house, a white Ben Hogan golf cap whose little button was missing from the front and a fancy gentleman's cane, painted white years ago for the Prom. On his tanned and handsome face he sported a pair of Benjamin Franklin glasses which he held together at one hinge with a few strips of Scotch tape. His father Irby's gold watch protruded from his vest pocket with its chain dangling down. He found it on his mother's bedroom dresser; it was the only relic his mother had saved for him. When he stood up after the service, I noticed he had on a white belt, too. It was truly delightful to gaze upon him in that outfit; he looked like a cross between Truman Capote and Colonel Sanders. I noticed he had on his old crimson socks which perfectly matched the handkerchief he had painstakingly folded into his suit pocket. He would have been immaculate had he remembered to bring some clean underpants and an undershirt.

Uncle Toogie could hardly contain himself during the wait for the ceremony to begin or during it. Jane and I were sitting directly behind him, waiting for him to explode and squeeze the life out of his nephew. Toogie told me after the service he thought Pete had staged his late entrance and that crazy outfit on purpose; he said he did that solely to disrespect his mother on the day she was buried. Toogie said that was exactly how he had remembered that sorry loser, his nephew. I preferred to believe Pete had just picked the wrong time to get high.

Pete slowly settled into the seat reserved for him, but before the preacher could begin, the sound of a song drifted up the hill from below us. The music circled around us counter-clockwise as a yellow Radio Cab with its radio turned to full volume approached the burial party. The cabbie continued up the single-lane road to a spot on the hill about twenty yards from the gravesite and slammed on the brakes; someone was in a hurry to join us. The music stopped, and a woman in her late thirties jumped out.

I whispered in Jane's ear, "Dang; I liked that song. *The Green, Green Grass of Home*... by Tom Jones. Ask 'em to play it again. Did you hear the words? They're perfect. *Yes, they'll all come to see me, as they lay me 'neath the green, green grass of home.* Play it at my funeral."

"Shut up, dammit; you're at a funeral. That was the cab radio; they can't play it again," she corrected me, as usual.

Everyone under the tent stopped their fidgeting as the scene unfolded above and in front of us. The whole funeral party stared in wonder, frozen in thought. The late attendee whistled at Peter Boy, waved to the crowd and giggled. I thought, oh, my God. The woman said something to the cabbie, and he cut his engine; he was going to wait. She started down the little bank sloping to the grave, and Peter Boy carefully rose from his little metal chair and returned her wave and giggle. The neck and ears of Pete's uncle Toogie turned fire engine red. The woman was dressed in a one-piece jumpsuit completely colored bright orange. She quickly approached us, and as she passed the coffin, she stopped and turned to look at the flowers resting neatly on its top. Everyone gasped when they saw her jumpsuit had the capital letters **RCCI** spread in bold black letters completely across her back.

Every local person at the funeral knew **RCCI** stood for the Richmond County Correctional Institution, but a couple of out-of-town elderly folks were asking, "What's that writing on her back?" They clammed up as soon as they found out.

One of Pete's old girl friends from high school had been given a pass from jail to attend his mother's funeral at his invitation. Could that possibly be true? Surely, a law prevented anyone from getting out of jail to go to the funeral of somebody's mother: a friend's mother, no kin at all, no one at all, no politician, some absolute stranger. In that same situation, most people would choose to spend what little funds they had to bail themselves out of jail rather than pay for a cab all the way across town and back. I wondered if the jail had a new policy that allowed prisoners to wear their prison work uniforms 'off base' like we sailors were finally allowed to do. Hopefully, the inmates weren't allowed to wear their jumpsuits inside a nightclub to drink and smoke pot with their friends or liberated cellmates.

In an amorous mood, Pete had called his old acquaintance's house the morning after we got to Augusta. Her momma said she had been picked up on some rinky-dink corporate theft charges, undeniably a bogus case of mistaken identity, and was getting out of jail in a day or two. The girl's momma asked Pete to please go see her; she was all alone and needed a friend, and it would be the Christian thing to do. She was still the good girl he remembered from high school, she said. Pete indeed remembered the girl, that's why he called her; she was the most promiscuous female to have passed through those hallowed halls in thirty years.

Pete called me at my mother's that same morning, and we agreed we'd just meet at Saturday's graveside service unless something important came up. During the day, he made arrangements to meet with the funeral director and his mother's lawyer, then took a cab downtown and rented a car. He drove to the funeral home and the lawyer's office, where he called the jail, a little too late, to inquire about visitation times and rules. He went to the S & S for an early supper, and since it was Wednesday evening, and he had nothing better to do, he went to Horne's for a drink. The bar's only other customer approached him an hour later and struck up a conversation. After he checked Pete out, the man offered him some drugs, and they got high together.

The next morning, Pete got up at noon and showered. He decided to go to visitation and line up something for the night the girl was released. It was either that, or he would have to go downtown to the Georgian Hotel, or maybe the Whistle Stop. He had nothing to lose by trying, and neither did the girl. He drove over to the Huddle House and had a big breakfast with all the good stuff you can't get on a ship. Pancakes with whipped cream and strawberries, ham and pork sausage links, a three-egg Western omelet and a glass of ice cold milk. And lots of fresh, hot coffee. They had the best coffee at the Huddle House; they'd make it fresh just for you if you asked. By the time Pete had finished stuffing himself, visitation was over. He'd have to try it tomorrow, he guessed. He went downtown and walked around Broad

Street looking at all the girls and shops; it was just like being on liberty without any other sailors around. Everybody commented on his Navy uniform, and he smiled in appreciation and said thanks. He watched the hit movie *Midnight Cowboy* at the Imperial Theater and then went to see *Butch Cassidy and the Sundance Kid* at the Miller. He was wired from all the sound and fury and needed a drink to calm himself down, so he headed back to Horne's as the sun set. After a while, the man from the previous night sat down at his table but made Pete pay for his drugs that time.

Pete got up at ten on Friday and made it to visitation on time. The girl's public defender also dropped by to talk to her, and Pete explained about the funeral and asked if the girl could be freed by Saturday. The defender thought that this girl was actually innocent, although most of his clients weren't, and was sympathetic to her plight. He said he would discuss the discrepancies in the case with the judge, and he did, in the judge's chambers over a ham sandwich and a cup of soup. Deciding to give the girl a break, the judge called the jailer himself and said he was sending some documents over to the jail releasing the girl for seventy-two hours on her own recognizance. The jailer told the girl, and she told Pete when he called on her at the jail. She promised Pete she would pick up where they had left off in high school and show him a great time. She said she knew where to get the best drugs money could buy, and she had plenty of money stashed away for a rainy day. Pete told her to be at the cemetery Saturday before four, and she said fine.

The judge put his signed documents and instructions on his secretary's desk and headed out to tee it up at his country club. The secretary had asked earlier to take her ninety-year-old mother to her oncologist, and he had forgotten. The next morning, Saturday, the secretary diligently ducked into the office to make sure everything for Friday was done. She saw the judge's papers and decided she'd take them by the jail on her way home, but first she had to pick out some new shoes for a party she was giving and get a prescription filled for her mother at Walgreen's. She got a quick bite to eat at Smoak's and then headed home. She didn't get to the jail until about three fifteen. Pete's girl, the inmate, only had time to sign out and call a cab. She didn't have time to change into her street clothes and, besides, she thought it would be kinda cute to go in her jail outfit; who does that? She knew Pete would like it; he had been so much fun in high school. The jailer agreed to let her go like that, but she had to come back and get her personal items and return her uniform after the funeral.

Jane stared at the girl, trying to figure out what the letters spelled and who the woman was. She asked me, "Is that woman a prostitute or something? Is she in jail?"

"You know she's in jail; who'd wear that if they weren't? How'd she get out?" I whispered, "Judge not. Whatever happened to love thy neighbor?"

"Oh, shut up. I'm not judging a damn soul."

None of the attendees were very charitable toward either Peter Boy or the woman with him, which was understandable considering the time and place. The problem was that almost all of the people at the funeral were his mother's friends, not his. Those folks had come to show their love and respect for her, not for her brother or Pete; they had come to say goodbye to her. They were her age and thought like she did, and all that foolishness at her funeral irritated the hell out of them. They had hardly heard of her brother or Pete, much less met them. No one there was Pete's friend, except the woman and me and Jane.

I helped physically restrain Toogie from attacking Pete at the service's end. But from my point of view, I found it hard to get too upset with Pete just because his friend came to the funeral; Pete had nothing sinister to do with that. The woman in orange had gone to a lot of trouble to be there. The pity of it was she had gotten there late and didn't have a place to sit next to him. Had they sat together it would have been like a University of Tennessee football game; one completely in orange and one completely in white. I tried not to judge her, or Pete, based on her one small lack of social decorum by wearing a prison jumpsuit to a funeral. Surely, it happens often. I tried to remember that Jesus associated with winos, addicts, thieves, and prostitutes every day.

When the initial shock of the appearance of Pete's friend finally died away, the preacher began the service. No one in the funeral party had said a word. Not a tear was shed during the entire service. No one mourned for Pete's mother; not her son nor her brother. The old adage proved true once again in its entirety: 'If in life one does not gladden, then in death one cannot sadden.' After the service, Pete looked for his friend and found her; she was quite easy to spot. They huddled together, and Jane and I were the only people who approached them or spoke to them. No one knew what to say or what to do, so they avoided them. I felt sorry for Pete. The funeral was over, and everybody departed. Jane and I told Pete to come to my mother's house when he felt like it and then walked to our car. The girl climbed up the hill to her cab and left.

Pete called me and asked if we could spend an extra night in Augusta. I said okay, but we had to be back at Oceana before Monday night at eight. He said he had a big night planned. I'm sure he did.

A month later, sitting at our little kitchen table in Oceana, Jane and I decided we wouldn't go home to see either of our families at Christmas. We were afraid something might happen that would keep me from reporting to the ship on schedule. Worried about running out of money, we had not sent anyone presents. We were troubled about not saying goodbye to our families, but we were anxious about the upcoming cruise. We'd never spent a night away from each other since we married in June. We decided to go to a movie one Saturday afternoon at the Princess Anne theatre. *Scrooge* with Albert Finney had just been released a few weeks earlier, and we wanted to see it. The time of year, our upcoming separation, and our guilt at not going home, brought tears to our eyes as we watched that old scoundrel Scrooge change into a human being. We both cried like babies seeing what a positive difference a change of heart can make in a person's life. That movie gave us the Christmas spirit. We immediately went out and purchased simple presents we could afford for everyone in our families. We wrapped and packed the Pittsburgh presents and mailed them as we headed out of town. We drove to Georgia to hand-deliver the other gifts we had bought. We had calculated just how much we could spend on gifts and travel to Georgia and leave Jane enough money to survive til next payday.

My brother and sister-in-law from Alabama had come over to say Merry Christmas and good luck on my cruise, and I asked her to keep me posted on her poppa's inside information on the Lt. Calley murder trial at Ft. Benning. Jane had picked me up a copy of the November 1970 Esquire magazine at the Post Exchange, and I was both flabbergasted and fascinated about the My Lai incident. Sue said she would. We kissed everybody goodbye and headed to Oceana. It was another wonderful Christmas.

Time was drawing to an end; the week arrived when we loaded all our stuff on the *Forrestal*. We were finally ready for our relatively short six-month cruise as a part of Carrier Air Wing 17 and Carrier Division Four. All we needed were our planes and aviators to be at full complement. About half of our aircraft would be on board when we pulled out of Pier 12 in Norfolk. We would pick up the jets of three more squadrons when we stopped at NAS Mayport to load them aboard.

On this her 9th Mediterranean deployment, the *Forrestal* would carry nine aircraft squadrons with seven different types of planes and two types of helicopters. Squadrons VF-11 and VF-74 flew F4B Phantoms. Squadrons VA-81 and VA-83 were made up of A7E Corsairs. Our VA-85 squadron had A6A and KA6D Intruders, and RVAH-7 flew RA5C Vigilantes. The Marine Corps squadron VMCJ-2 DET 59 was on board, flying EA6As for electronic countermeasure missions. VAW-126 flew the E2b Hawkeye Airborne Early Warning twin-turboprop loaded with radar to detect approaching enemy jets or missiles. Helicopter Anti-Submarine squadron HS-3 with their Trident and Sea King helicopters completed our deployment aircraft requirements.

16. THE RESCUE OF THE FLAMINGO

Early on the morning of January 5, 1971, Pete and I once again stood on a pier and stretched our necks to gawk bug-eyed at our new home. Our squadron's equipment had already been stowed away; all we had to do was climb up the gang plank and find our berths. We were very hung over, and I just wanted to climb into my rack and die. Jane had driven us to the Norfolk pier to bid us farewell. She was such a lovely girl and a beautiful wife, and I knew I was really going to miss her. We didn't drag it out; we kissed and said I love you, and I was gone. The excitement of the upcoming cruise with all its historic ports of call did not lessen the pain of my departure, nor my hangover. But I quickly adjusted to carrier duty. I covered my pillow and sack with the pillowcase and bed cover we'd been given and pulled the heavy dark-green plastic cover closed at the front of my rack. I was asleep before the ship got underway.

It seemed like I was going to waste away half my life sleeping it off after a night of heavy drinking. But I remembered that my mother and father were both very heavy drinkers, certainly alcoholics, so Jane and I had decided early in our marriage to slow down.

My daddy used to drink bourbon and Coke every night from an iced tea glass, and momma joined him half the time. Most times they had more than one. One typical afternoon after I returned from our hangar at Oceana, Jane fixed us an iced tea glass filled to the brim with vodka and grapefruit juice and a lot of cold, refreshing ice. Halfway through, we looked at each other, and I said, "This is the third night in a row that we've had these big drinks, and it's not even the weekend." From that point on, we only indulged in too much booze during a celebration or the arrival or departure of something. But I never asked Jane what she drank, or what she did, while I was gone, and she never asked me.

We were steaming on the high seas when I finally felt like getting out of my rack. You had to report for duty and all that when you're in the military, so I asked a friend to tell the Chief that I was sick. The Chief knew what sickness I had and left me alone. Charlie Cranston was a good man. Every sailor in our AQ shop loved him. I know I did; he had already brought me back from the border patrol in Tia Juana and the Shore Patrol in Subic Bay. Poor man, what were Pete and I going to do to him next?

Most Navy Chiefs (E7) and First Class Petty Officers (E6) got along very well with the men under their control. That was only natural because those older men had to have reenlisted to be eligible for those rates. The Navy has rates, and the Army has ranks. Officers are Os, and enlisted men are Es. At that time, in 1971, a sailor could go no higher than the rate of a Second Class Petty Officer (E5) if he only stayed in four years, and that's what Pete and I had been for a year. Our Chiefs and First Class Petty Officers had been associating with young, dumb sailors, virtual kids like us, day in and day out, year after year, for at least five years. They lived at sea beside their men from six to nine months at a stretch in a metal compartment called a shop about twenty feet wide by thirty feet long. When all our gear and several computer parts were in our shop with six or seven sailors, a man could hardly turn around. And everyone was chain-smoking and drinking coffee. That's why the older sailors got along so well with the younger sailors; they were always around them, they were a lot like them, they knew what was on their minds, and they could defuse them before they exploded. Like the Lost Boys in Neverland, they never grew old.

In port, the older sailors usually stuck to themselves. That was probably because they were tired of us being stuck up their butts all the time, or maybe they didn't want to have to babysit us when we got commode-hugging drunk. Of course, we newer guys wanted to be with our own friends, too, whenever we went on liberty. But if we saw a Chief or a First Class from our squadron whom we liked, we immediately clutched onto him. We let our hair down because we were excited to see those guys ashore, and if they had been drinking and were happy to see us, we usually hung out with them a while. They had the greatest stories to tell; stories of wild parties, terrible binges, chaos at sea. Their stories usually couldn't be told in mixed company, but their stories were hilarious and couldn't have been made up. The camaraderie we shared with them was never forgotten.

The next morning, I stood on the fantail at the very back of the ship with Peter Boy by my side. The ship's wake churned the Atlantic into a beautiful corridor of light green saltwater eighty yards wide and two miles long. It looked like someone had poured millions of gallons of mouthwash into the ocean directly behind the ship. Under the surface, the ship's four huge props, each with four massive blades, had turned that water over and over until enough air had been infused into it to turn it into a giant bubbly and frothy path. I looked up from the ship's wake and saw the sun was off to port, the left side of the ship. The time of day was still morning.

"Pete, I thought we were heading east across the Atlantic. Why is the sun on our left? Why isn't it up front?"

"Yeah, it should be, but I don't know why it isn't. We're heading south."

Our Captain received orders to sail to the U.S. Virgin Islands before we crossed to the Mediterranean. Our islands are officially called the Virgin Islands of the United States to differentiate them from their northeastern neighbor, the British territory of the Virgin Islands. The *Forrestal* had completed her operation readiness inspection making sure everything was ship-shape and then anchored off the coast of St. Thomas. We were disappointed to find that we couldn't go ashore, so we spent the remainder of the day looking out at the beautiful scenery and fine-tuning our planes. The next day, the *Forrestal* sailed east until we arrived at Cadiz on the west coast of Spain at the Spanish Naval Station Rota. We relieved the U.S.S. *Independence* (CV-62) on January 24th in a change of command ceremony. This port call sadly ended with the same results for us, the little men of the ship, for there was no liberty. We sailed through the Strait of Gibraltar that night.

This cruise throughout the Mediterranean was primarily staged to allow the naval vessels of our fleet to be seen by our allies and our adversaries. We were on display, on a diplomatic mission, not a combat mission. The diplomacy came from the Teddy Roosevelt era; speak softly and carry a big stick. An aircraft carrier qualifies as a big stick in my book. We were there to show the world once again that we could kick some butt if the situation called for it. We were going to sail around the Med taking in the sights and making our presence known for all the world to see. It was going to be a pleasure cruise, if everything went our way.

The *Forrestal* was fully loaded with armed missiles and bombs, in case they were needed. We trained and practiced and showed off to visiting dignitaries and newsmen. We were on a mission to show the public, the people of the Mediterranean and far beyond, how mighty but benevolent we were. We went to port more often than our pocketbooks could handle. We drank and played on leave and worked and fretted on the ship. We hardly had time to get back to work at sea before we had to get ready to head for another port. The difference between being in the Mediterranean at that time versus being in the Western Pacific was like practicing for a Super Bowl game instead of actually playing in the game. We were kinda playing on Easy Street in the Med, while our boys on the other side of the globe were dying in record numbers. However, it was all a necessary part of our government's plan, giving them the benefit of the doubt that they had one. We were preventing any aggressors in that part of the world from thinking they could start something unholy while we were fully occupied with Vietnam.

Pete and I were still best buddies, but Jane had put a major dent in our relationship. I just naturally preferred her company to his. It was a biological thing, not an emotional thing. During the time I had been with her, Pete had met some new friends and had taken on some new hobbies. We had both changed, and I felt like he was using drugs; I hoped it was only a little marijuana. It was just the way he talked and the things he was interested in, and I could have been wrong. I'd just have to wait and see. I'd have to watch him on liberty.

In late January, the *Forrestal* anchored off the seawall of Valletta, Malta, on our first visit to a port on this cruise. The weather was so stormy and the sea so choppy that liberty was cancelled, and our ship visitation for important foreign dignitaries was rescheduled. We waited for the weather to clear, but it kept storming, and after a couple of days we pushed off.

My 1970 W-2 came, and my earnings were $4,138 for the year. My take-home pay had not increased because I was repaying my college and car loans. Jane and I didn't have to file a tax return because half of my pay was tax-free combat pay, and she had taught elementary school in Pittsburgh only for the first half of the year. She earned $2,230 for that and didn't work for wages while we were in Oceana.

The *Forrestal* began prowling the Ionian Sea west of Greece in early February. She received word that the *Flamingo*, a Panamanian-flagged ore ship, had lost power and was foundering off the southern coast of

Italy. On February 7th the *Flamingo* drifted at the mercy of the wind and tide as she was taking on water and sinking. Several of our carrier group's destroyers went to help but couldn't save her; she was so much larger than they were, and she was uncontrollable. Four SH-3D Sea Kings from squadron HS-3 were sent in winds exceeding 60 knots to rescue the crew, if it were possible. We watched the drama live on the ship's television as the enormous seas washed upon our flight deck, making the whole carrier shutter. We sat almost horror-stricken in our AQ shop waiting for the typhoon or hurricane, or whatever it was, to stop. We sat there watching the file cabinet drawers on each side of our shop open and close as the ship rolled and rolled, back and forth. And we were on an aircraft carrier.

"Whatta doing sitting there looking like the world's coming to an end?" the Chief asked. "This isn't crap. These ships are filled with oil and supplies all below decks. They'd break in half at the waterline before they'd flop over. Are you serious? Think of it this way: it's like being on a plane; once you get on it, you might as well forget about anything bad happening. You're not going to wimp out like some psycho and tell 'em you want to get off. Who'd do that? It's the same on a ship; who in the hell would say they need to get off? Oh, I know some section 8s do stuff like that, but not normal, realistic people. Once you get on the ship, relax, there's nothing you can do about anything."

The hurricane or cyclone, sometimes called a Medicane by the locals, worried me because I had looked at our aircraft carriers pretty carefully over the last two years and noticed the enormous amount of steel and aircraft and equipment exposed above the sea. A carrier's bottom near the waterline is a skinny little thing compared to the gigantic amount of ship sitting high above that. I worried if the ship could keep from toppling over into the ocean in a typhoon. I felt the same way every time we turned into a strong wind while cruising at full speed during flight ops. The tilt of the carrier caused by the centrifugal force of the turn and the furious wind pushing on her side during the turn made me think of my afterlife, but it was necessary to turn the ship into the wind to maximize the wind speed over the jets' wings when they were launched, whether it bothered me or not.

The four HS-3 copters returned to the ship with all twenty of the crewmen and passengers of the *Flamingo*. The survivors hopped off the copters and walked through the hangar deck. One of them was clearly shown on our flight deck camera as being a woman. Our sailors fed them and gave them all the medical attention they needed and flew them to Naval Air Facility Sigonella on Sicily the next day when the weather improved. We presumed the *Flamingo* was lost at sea; we never heard.

In early February, we were anchored at Saint Paul's Bay in Malta. The town was the largest town in the northern part of the country and was named for the place where Saint Paul shipwrecked on his way from Caesarea to Rome, as reported in the Acts of the Apostles. The Secretary of the Navy and the Commander of the Sixth Fleet joined us there to present medals to the sixteen men who participated in the rescue of the *Flamingo*'s crew. It was a glorious day for us, because not only were our shipmates honored, but Secretary Chaffee announced that we were the first participants in a new Navy strategy. As of that ceremony, the Navy had decided to let all of us start wearing civilian clothes when we went ashore. Previously, only Officers, Chiefs and First Class Petty Officers were allowed to do so, but now we all could. Had we been told that a month earlier, we could have brought some with us, instead of having to buy a new wardrobe on our skimpy wages.

We next cruised throughout the Ionian Sea and returned once again to the Maltese seawall of Valletta on the 22nd. Pete and I planned on going ashore to celebrate my 24th birthday. The weather was still terrible, and the Captain said all visitation between the ship and Valletta was restricted. We stayed on the ship, working and waiting, until the weather abated. Two evenings later, Pete and I were in a bar downtown throwing darts when it was time to get back to the ship for the night. We didn't have passes to stay ashore overnight. Pete said, "Screw it; let's just stay here. I'm tired of going back to the ship every damn night we're on liberty. They know we're okay; this is so mickey-mouse."

I was tired of all the rules myself, so I decided to stay with him. A shipmate drinking with us in the bar had a room in a hotel and let us sleep on his chairs. We were absent at roll call the next morning, but some of our fellow AQs told the boss where we were. Chief Cranston once again came to get Pete and me. We were hungry, hung-over, and disgusting. Our punishment for being absent without official leave, casually referred to as being AWOL, was having to miss two days of liberty in Athens, Greece. Believe me, if we had pulled that AWOL crap on a West Pac deployment, I'd still be in the brig, but some slack is usually given for mental lapses and stupidity when times are not as desperate as on a combat cruise. Pete wondered if that same casualness and inattention to detail contributed to the errors or mental lapses that led to the *Forrestal*'s deadly fire of '67 on just the fifth day of her first combat deployment.

Soon after our departure from Malta, we worked for about a month patrolling the eastern side of Greece in the Aegean Sea while holding flight ops the entire time. Our government wanted Turkey and the Communist countries called the Eastern Bloc to see what we were capable of doing if we had to calm the region in the event of a conflict.

We arrived in Athens in the middle of March of '71 and were given liberty. We had anchored at Athens' port of Piraeus. Pete and I were particularly interested in the Acropolis, the small mountain of rocks in the center of town on which sits the magnificent ruins of the Parthenon. There were also many theaters and museums to see, but they all cost money. A Mediterranean cruise like the one we were on could only be fully enjoyed and appreciated by someone with a lot of money, which did not include most first-tour men of the United States Navy. Especially without combat pay.

Pete summed up our situation for me one day. "It's exactly like being back home. Over here we spend all our money buying cigarettes on the ship and beer and wine on shore. So all our pay goes to smoking and drinking. We don't have any money to do something special because we spend it all first. We need to stop smoking and drinking or find something that pays more money."

"So how in the hell does that have anything to do with back home?" I asked.

"It's simple. All those folks who never have any money for food or housing back home always have money for smoking and drinking. If they'd cut out the smoking and drinking, they'd be able to afford food and clothing for their kids and a roof over their heads."

"Pete, you never let up. How'd you get so mean?"

"How'd you get so stupid?" he replied.

Pete and I were on restriction for two days in Athens due to our failure to report back to the ship in Malta. Thank God. We couldn't have stood those extra two days on liberty anyway; we were running low on cash, as usual, but we still had plenty of cigarettes. The first thing smokers did on payday was buy enough cigarettes to last until the next payday. Whenever a sailor was asked if he had an extra cigarette, he always searched through his pack, trying to find that extra one, and then said, "An extra cigarette? Sorry, this pack only came with twenty."

It wasn't a matter of stinginess; it was just contempt. Someone who lacked the self-discipline to buy enough cigarettes to last until next payday was an imbecile unworthy of most sailors' sympathy. Pete, a man who considered himself an expert on self-discipline, said that. It was ironic that such a ridiculous statement came from his lips because Pete and I displayed the least amount of self-discipline of any sailors we knew.

17. CHAVA LASKA'S HOME ARMY

The epitome of positive self-discipline, something that required more inner-strength and self-control than Pete and I could muster, was Aviation Fire Control Technician First Class Petty Officer Jerzy Laska. Jerzy, the Polish name for George, went by Jerry and sometimes answered to our nickname Lash. He thought the name Jerry was more macho than Gerry or George. Absolutely the second most respected man behind Chief Cranston in our outfit, Jerry was the shop boss whenever the Chief was gone. A top-notch, hard-headed, tough old salt from Missoula, Montana, he was a nitpicker for detail and self-discipline. He would have been a real-life cowboy if he hadn't loved jet fighters so much. He tolerated Pete and me only because we liked to hear the stories about his family and the West that he loved to tell. He hated us for our laid-back, sloppy attitudes. He would have kicked us out of his Navy if he had the chance, but the Chief always intervened for us. He often told us of his experiences on Yankee Station and Bangkok and Olongapo, but we were also interested in his stories of the Jews in his family, especially during World War II, whenever he would tell one.

Sitting in the shop while everybody else was on liberty in Athens, Pete and I were bored to death. We had to kill the first two days in the shop for that little incident we had in Valletta when we 'forgot' to get back to the ship for muster.

"Lash," I said, trying to get a rise out of him, "Pete and I want you to tell us some stories."

"Why aren't you boys in town? Oh, yeah, now I remember," he said sarcastically. "You guys are an embarrassment to my Navy. Why don't you quit and go home? I'll fill out some forms for you right now and get an officer to approve them."

"Come on, Jerry, give us a break. We know we cause you and the Chief a lot of trouble, but we're sorry, and we'll quit screwing up; you'll see. We just get all wound up, or sometimes a little too drunk, but that's what liberty's for, isn't it? Come on, do us a favor, man, we've got nothing to do; tell us another story about your family. The Polish Jews in World War II."

"You want to hear about my Hebrews? You sure? Well, I can stand it if you can, but I can't remember where we started or stopped before. You may have already heard some of this, so stop me if you want," he said.

"I'll start with the beginning. So you'll have a little background…My father moved his entire Jewish family from Lithuania to Poland in 1920, right after World War I. He was doing well and wasn't in a hurry to get us out of Europe, until it looked like war was coming. He said, 'Poland can defeat any invaders, including the Germans.' But he was wrong; we were in grave danger. To escape the imminent holocaust, my father decided to move us to the States in June of '39. We were devout Jews, but we didn't have any trouble getting out because my father worked for the U. S. Embassy in Warsaw. He just applied for a permit to go to New York to see the World's Fair.

"My mother crammed four suitcases with as much as they could hold, and each of us carried a tote bag loaded to the brim. We took nothing but our clothes and jewelry. We tried to look like four innocent tourists heading to New York. We kids weren't told we wouldn't be returning. All of us appeared a little breathless, but anybody going overseas to the Big Apple for the first time would be. The story fit like a glove. We got to New York and were given political asylum."

"Thank God for the USA," Pete said.

"Thank God, indeed. Uncle Daniel, my father's brother, didn't have a state or an international job and waited too late to escape. There wasn't going to be any asylum or any future for his family, except for one daughter, Chava, or Life in Yiddish.

"In the Second World War, Warsaw was the scene of two different uprisings against the Nazis; one by the Jews of its Ghetto in 1943 and the other by the Polish underground in 1944. Adolph Hitler had risen to power because he promised the Germans two things they didn't have in the '30s: jobs and food.

"Under the Molotov-Ribbentrop Pact of August 1939, Russia and Germany agreed to divide Poland between themselves from north to south. The River Bug was the boundary between the German half on the west, and the Russian half on the east. The Daniel Laska family was east in the Russian zone on holiday when the Germans unexpectedly attacked in September. The border of the Russian-German zones was closed, and the Laska's home was west in the German zone."

Jerry lit a cigarette. "Okay. My cousin Chava told me after the war that her family could only get home by secretly crossing the River Bug. Daniel paid two men to guide them across to the German zone, and they waited in the woods. They joined a few other refugees that evening and after a couple of hours of walking, they were almost there."

Jerry rubbed his forehead and said, "During the night, a young baby started crying, and her parents couldn't quieten her. One old man said to smother her. As everyone dreaded, the baby gave them away. Within minutes they were staring at four Russian border guards who demanded to know what they were doing there. Nobody said a word until one man yelled that they had just escaped from the German zone. Of course, that was a lie; that's where they were heading."

"Quick thinking," Pete said.

"It saved the day. The proud guards had stopped the escape and were told to send the escapees back to the German zone. They were escorted to the river's edge and were rowed safely to the other bank to German guards. In the morning, the Germans marched them to a railway station. The family, cold and hungry, was forced onto a freight train. Two days later, never fed, they arrived in Warsaw.

"Chava's family lived in a large apartment building in the ritzy part of Warsaw. They were ordered to move into the Jewish Ghetto on two days' notice with one suitcase each, forcing them to leave their valuable furniture and artwork behind. Nothing was sold; the Poles knew the Jews didn't have a market for their stuff, so they made them leave it and later sold it themselves. The Poles were just like all the other people; they screwed the Jews."

"That'll always be true, and probably by everybody," said Pete.

"The largest Jewish uprising of the war occurred in the Warsaw ghetto. The first resistance to the Nazis started in April of '43 when the last of the Warsaw Jews refused to go to the concentration camp at Treblinka. In small groups, the Jews fought the SS throughout the city, until they died. The SS Brigadefuhrer received orders to burn every building in the ghetto. The block by block destruction ended in May, and the entire ghetto was gone. Thirteen thousand Jews were killed; half were burned alive or were suffocated. The dead included Daniel Laska, my aunt and three of my four cousins.

"My cousin Chava had escaped from the ghetto and joined the Polish Home Army two months earlier passing herself off as a Catholic, but I guess they would have taken anyone," Jerry said. "She said she'll agonize for the rest of her life that she didn't die with her family. She hated that very few Poles lifted a finger to help her Jews.

"Chava and several other Home Army recruits lived in an abandoned basement as the fighting got closer. Told that a nearby street needed a barricade to slow some German tanks, they ran to help. By the time they finished, the Germans were there, so they took off. They reached a neighborhood that was in friendly hands, where one of her friends lived. They were safe but had to report to their duty stations. Chava's friend had not joined the resistance but wanted to be a part of the uprising and asked if she could go with them. She told her parents goodbye, and that was the last time she saw her mother or father.

"They ran through basements and down streets hiding behind barricades. They reached an exposed intersection they had to cross, but a sniper was shooting at anyone brave enough to try it. They saw one man get shot, and another was lying dead in the street. Chava's friend got upset and wanted to go home; she wasn't tough like Chava and had seen enough blood. Chava wouldn't quit because she had to obey her orders. Her friend stood up to say goodbye, stuck her head over the barricade to see if it was safe to run, and sank back down; the sniper had popped the back of her head off as if it had been a ripe watermelon."

Pete said, "God, that's sickening. They were just a bunch of kids."

"Younger than us," I added.

Jerry kept on, "The resistance was told to start the uprising when they heard the first shots of the Russian army on Warsaw's outskirts. The date was set for August of '44. The Polish government in exile in England had gotten the Russians to agree to support the resistance by moving into the city and forcing the Nazis out once the uprising began. All the Home Army had to do was hold off the Nazis for three days and then take cover, but soon after they began, the Russian guns quit firing. Chava and her fellow fighters had come out of hiding and were fully exposed and knew right then there'd be no help.

"Joseph Stalin never planned on helping the Home Army. He wanted Poland for himself and was willing to let the Germans keep Warsaw. He knew that just as soon as Hitler was defeated, which he knew was a certainty, he would pick it up like a discarded leftover.

"When he heard of the uprising, Nazi SS Himmler swore that Warsaw and its entire population would be destroyed as an example to every freedom fighter in Europe. The Nazis rounded up and executed all forty thousand civilians living in the fashionable Wola District."

"How can anyone ever forgive those animals?" I said.

"Not me... The resistance was forced into hiding but still attacked the Nazis whenever they could. Chava and her friends were hunted down like the wild animals they had become, through sewers filled with rotting humans and floating feces in the stinking cesspools under the city, until they were captured one by one and tortured and shot. The Germans cut off all water and electricity, and there was no food of any kind. Chava said the remaining Poles in Warsaw ate every animal in the city, even the rats. The people died at the rate of two thousand every day, and it was just a matter of time before the Home Army had to surrender. Chava said the only silver lining in that blaze of death was that her proud, confident father wasn't alive to see it."

"Daniel was her father?" Pete asked.

"Yeah, he was her father, my dad's brother, the one who knew they'd be safe in Poland. The one who asked what evil could happen to the Jews; they were in Poland, not Germany.

"Anyway... Chava was one of twelve hundred girls who were loaded onto cattle cars with other fighters to be shipped off to a prisoner of war camp. They were joined by thousands of Russians. Chava was humiliated by the Germans, but she knows she was lucky she wasn't raped and beaten. The Germans weren't set up to handle women prisoners, so they processed them like men. Every two weeks the women were made to strip naked, in front of everybody, while their clothes and hair were deloused... Were deloused or was deloused? Whichever; I can't keep those stupid rules straight."

"Don't worry about it, Lash; you're a First Class Petty Officer, not a damn English teacher," I said.

"You're right, egghead, for once. Okay, the Germans made the female Russians do the delousing because of the unbearable stench. After two months the women were moved to a camp near the Elbe where they joined six thousand other prisoners. Some had been there five years. The Nazis let them keep the Red Cross clothing and food in their care packages. Many of the guards had relatives over here as prisoners, and they expected the Americans to feed them, too."

"We did; we had a POW camp in Augusta. Momma said the prisoners would get lost on work details and walk all through the neighborhoods before some housewife would call the camp and tell 'em to come get 'em before they got in trouble. They were like kids, but you know they were in a better place over here. I doubt if a single one would have ever gone back."

"Interesting …Well, the packages also had chocolates and cigarettes which they used for bartering. Chava told me more than once that without a doubt the stinking Russians would have kept every piece for themselves. Her daily diet was a half-pound of bread with margarine and jelly which someone made from moldy beets or carrots. She learned to cut the bread very thin, so that it looked like she had a lot. Her dream, she said, was to have enough bread for two days at one time, but it never came true."

"Have the frigging Russians ever been worth a crap to anybody?" Pete asked.

"Early in December of '44, Chava was moved and made to work as a slave in an ammo factory, which the Nazis routinely did in violation of the Geneva Convention. In February, she worked three turning machines, always standing, from six to six. In April of '45, bombs started falling near the factories, and each time, the workers were sent into a nearby field. The Americans were bombing the Germans from the west, and the Russians were bombing them from the east. The prisoners prayed the Americans would come first; they knew that even as bad as things were, they would be worse under the Russians. They were forced to keep up their regular work, even though they were up half the night avoiding bombs.

"One day the factory manager called the prisoners together and gave them an impressive speech about how fair the Germans had been. They were then split into ten groups of about one hundred and twenty each and sent to another camp. Each group during the trip was guarded by only six soldiers. The first night came after a seven mile march. Chava's group found an old abandoned barn. A few of the girls escaped during the night but most of them stayed in the barn because it was too cold outside. They didn't have any warm clothing, so they just huddled together.

"In the morning, they refused to go any farther. The guards tried to make them get up and start marching. They threatened them but finally gave up. Instead of shooting them, the Germans actually stayed with them for their protection for two days until the Americans found them, unharmed, in April of '45. Chava was freed and was sent to a civilian reclamation camp.

"In their two-month struggle, eighteen thousand Polish Home Army fighters died in Warsaw, and twelve thousand were wounded. Two hundred and fifty thousand civilians were killed during the Uprising, and the Germans suffered ten thousand soldiers killed and another ten thousand wounded.

"My cousin Chava survived and was welcomed to America by my father three years after the war ended," Jerry said.

"She had a lot of balls, Lash. I guess y'all got it from the same DNA. Good stock. You know I've only known five or six Jews personally; all in school. I was in the advanced class with all the smart people, and of course, all the Jews were in there. I don't know if they were all smart, or maybe their parents always made them do their homework. Whatever, but now I've met you and a Marine instructor we had in Millington. Good people. Thanks, man," I said.

On a cold night at the end of March, we left Athens and cruised through the Strait of Messina into the Tyrrhenian Sea. A damaged A6 was to be unloaded and transported to NAS Naples for repair, but high seas made it too dangerous, so we anchored in the Bay of Naples until the next day when the weather improved.

Over the Easter holidays in '71, the Navy scheduled a dependents' flight to Barcelona so some of our wives could join us for ten days. Jane signed up for the flight with her girlfriends from their Wives Club at Oceana and flew from Oceana to Barcelona. We anchored near Barcelona, and she was waiting for me at the pier when we walked down the gangplank of the ferry boat the Navy hired to take us ashore. She was a sight for sore eyes. She had cut her beautiful long hair and now sported a sexy-looking shaggy bob. I took two week's leave to safely cover the entire ten days she was there. Pete got his feelings hurt and went off somewhere with a couple of roughnecks.

My good Miami friend Gabe, the boy with the French mother and the Basque father, had an aunt and uncle living in Barcelona. The aunt was a housewife, and the uncle was a rich businessman who had been elected President of the Bulls at the Plaza de Toros Monumental arena. Jane and I were invited to their downtown penthouse with Gabe, of course, as often as we wished.

Each morning Jane and I walked down La Rambla, Barcelona's forty-yard wide, tree-covered boulevard that stretched for three quarters of a mile to the Mediterranean. We started each day with a stroll ending at the statue of Christopher Columbus overlooking the harbor where he had set sail for the New World. The replica of the Santa Maria was tied to the dock while we were there. We ate Spanish food and listened to the clatter of high heels and metal taps slamming the floor while the Gypsy-looking women danced the Flamenco. We were astounded by the Church of the Sagrada Familia built by local architect Gaudi; apparently the inspiration for the word "gaudy." The church was begun in 1882 and due to illness, death, civil war, world wars, depressions and recessions is estimated to be completed in 2023.

From the first row in the middle of the field, we watched a professional women's soccer match with Gabe's family. We made hand signs to Snowflake, the world's only albino gorilla. We cheered for the bulls during the fights on Easter Sunday at the Monumental bullring as we sat in the president's box. Europeans didn't boo when they didn't like something; they whistled. So Jane and I whistled whenever the bulls were being worn down and stabbed to death. The Spaniards knew we were just ignorant Americans who didn't know any better, and thank God, Gabe's uncle wasn't there. Jane and I eventually decided the pageantry and showmanship of the bullfights was justified, knowing that all the meat from the slaughtered bulls was given to the local orphanages in the Barcelona area.

The wonderful thing about being in a foreign country is experiencing its many different customs and traditions. On La Rambla, we laughed as we watched children tinkle on the edge of the street. Parents held their little girls off the ground by their tiny arms, and the girls put their feet onto the trees planted long ago in sculptured holes in the sidewalk. One parent pulled her clothes aside and held them out of harm's way. The young lady then bent her knees and urinated in a stream flowing down to the base of the tree. Job done, she hopped down and straightened her panties and dress. The young boys just dropped their pants and underpants right there by a tree and leaned toward it and urinated like dogs at a fireplug, but they didn't lift their legs. No harm, no foul, that's the way it was done. In Palma de Majorca, I saw a woman urinate against a tree in the park. Her underbody was completely covered by her dress, but you could easily see the stream she let flow. Apparently, everyone did it; it was the norm, so easy and convenient. To them, the crazy people were the ones who didn't like it.

Another norm in Europe that was not quite the norm in the States was the passing of gas in public. No attention was paid to someone, man, woman or child, pooting in public in Europe. On several occasions, Pete and I experienced women at tavern tables next to ours, leaning to one side and audibly expelling gas with or without a smelly olfactory response. In mid-sentence in a group of students or vacationers or stewardesses, women leaned over and pooted without missing a syllable or looking over their shoulders to see if someone had noticed it. Pete said they were usually Germans, but I didn't know how he knew. Other olfactory issues arose when we encountered some people who didn't use any deodorant; many Germans, Indians, Mexicans, Pakistanis, Muslims, and Africans, to name a few, grow up in cultural environments that consider it a waste of time and money. Most folks need it; a few don't.

Time came for our hometown ladies to head back to the States, and I saw Jane off on a Thursday with the usual hugs and kisses and expressions of undying, never-ending love. It had been wonderful; she was so sweet and loving, and I knew how much I'd miss her...in time. The next day, on Friday afternoon, Gabe's family took me with them to their summer home on Spain's Costa Bravo where we were wined and dined over the weekend. With heavy hearts we eventually returned to the *Forrestal* which had been operating around or anchored off Barcelona those last two weeks. We pushed off to begin another week of flight ops in preparation for an upcoming NATO event. While we were cruising, I read the Newsweek magazine for March 8, 1971, that Jane had brought me to keep me abreast of the Calley trial.

On the 27th of April, we anchored at Majorca, the largest of Spain's Balearic Islands, near Palma de Majorca, the island's capital. Because of its legendary beauty, it was an extremely popular vacation spot particularly for Germans and Brits, but they didn't say vacation over there, they said holiday. The Phoenicians arrived in Majorca in the 8th century BC and put it under the control of Carthage in North Africa. After the 2nd Punic War, the island fell under the thumb of Rome. Nothing of note has happened there since then; however, twenty-six million tourists a year may disagree with that assessment. But on the other hand, that may be exactly why they go there.

Pete and I took the liberty launch into Palma and looked around. We walked along lovely walled yards bordering the main thoroughfare leading into town from the pier. The walls were covered with overhanging, droopy gardenia-like flowers and vines that smelled like expensive French perfume. A mile from the pier, we turned down a side street and then turned again back toward the pier. We strolled along the boardwalk looking at the yachts. We drank wine and ate cheese and crackers. We talked to two quiet young ladies who weren't pros. They were just two college kids on holiday from Germany, and we bought them a drink which soon warmed them up. Not long after we met them, Pete leaned over and whispered something in the pretty blonde's ear, and she smiled and nodded her head. He looked at me with a grin and said they'd see us later. I couldn't figure out how he had gotten to the girl so quickly. The other girl and I just sat there looking out at the harbor, enjoying the view. We had another drink; she bought that one. She spoke English very well, and we talked politics a while. We were running out of ideas when Pete and the other girl walked up. I had assumed they had gone somewhere to make love, maybe to her hotel room, but I could see by their eyes that she had taken him to get some drugs.

I wasn't a prude, but I didn't like that. Pete was my friend, but I was worried about being around him and any craziness he may cause when on drugs. He was already crazy as hell; imagine how he was on drugs. We only had four more months to serve, and I sure didn't want to get kicked out. I didn't care what Pete wanted; I wanted to finish my Navy obligation, stay out of trouble, and get my three years of college paid through the G.I. Bill, as promised. That's why I joined: three years of $4,400 each to complete my college education, thanks to

Uncle Sam, which means the American taxpayers, which means about half of America.

The four of us sat on the patio of the open-air restaurant near a crowded pier and drank a few more beers and smoked a couple of cigarettes. Pete and the blonde had had their fun and were rejoining the conversation. Nothing was going to happen with the girls except talk. That was all I expected, and that was all I wanted. Pete was looking for some more action, but with me and the other girl there, it wasn't going to happen. So solly, Joe; as they said in Sasebo. It came time to eat, and I was hungry. I didn't want to offer to buy the girls their supper, so I said goodbye and started back alone to the liberty launch to the ship. My day of sightseeing and chit-chatting in Majorca was over. I stayed on the ship and kept our shop operating the next day, and I never asked Pete what he did while I was at work. Pete and I had both been promoted to E5s a year earlier, and we took our turns working in port on regular schedules. E5 was as far as you were allowed to advance without reenlisting. So solly, Joe.

From April 28th until May 17th, we roamed all over the Mediterranean participating in Operation Dawn Patrol, a major air and sea exercise which included more than sixty ships and submarines and three hundred aircraft from all NATO countries. We completed those exercises and recovered our planes. Those impressive exercises meant nothing, except extra work, for the enlisted men on the *Forrestal*; they weren't announced and we weren't told; so we didn't even know we were participating in a so-called major NATO exercise. Life was just the same old, same old for us; the same stuff every day, just like a nine to five desk job in an accounting firm or an insurance office. But on our way to the next port, the Captain announced we were to be honored with an airshow in a half hour, and so we perked up and rushed to the flight deck to see it. The massive flyover of plane after plane was in celebration of the 20th anniversary of AFSOUTH, Allied Forces South Europe.

Our preparations for the NATO exercises were like those of any other day. If you worked the twelve-hour day shift, you got up around 0700 and roamed around and shot the shit until you finally went to the head to clean up and get squared away. After nature's call, you took a quick Navy shower, shaved and brushed your teeth. Back at your rack, you took a clean uniform from out of the storage area under the rack, or put on yesterday's clothes (or an earlier day's or a day or two before that, like when you were in college) if they were presentable. We AQBs of VA-85 worked in an enclosed room, our shop, on one side of the hangar deck and kept it fairly clean. We worked on the hangar deck and the flight deck on our jets, so we stayed pretty clean ourselves. The biggest obstacle to our looking ship shape was jet fuel, oil and exhaust gases from the planes. Black shoe sailors worked with grime, grease and oil below and above decks and were usually seen in oil stained, white work shirts, usually tee-shirts. Airedales worked around exhaust fumes and jet fuel on the flight deck all the time but could stay fairly clean. Bombs and missiles were everywhere, but they weren't dirty, and we stayed away from those sons-of-bitches anyway. Most sailors not in the belly of the ship had no major problem staying straight. Men who repaired electronic gear and aircraft and ship parts worked in special, air-conditioned shops off the hangar deck. My friends who repaired and tested our incoming systems and computer and radar parts were like any technicians you would see in any electronic shop in the States. On my off time, I would visit them to play a little bridge or read a Playboy magazine. Their schedules, unlike mine, weren't defined by flight operations. They worked at their own pace; I worked during my shift, or afterwards if something special had to be done. At sea we spent half our time physically working on our systems in the planes on the hangar deck, and half our time discussing in our shop what had to be done to them when they were recovered from flight ops.

Down to the chow hall after we squared away our racks, we would go stand in line while we waited for the entire U. S. Navy to be served ahead of us. The only thing I hated more than standing in a line was seeing some jackass breaking in line in front of me. On several occasions I went up and got in line right in front of him. I'd call my buddies to join me, and I'd lean against the wall, showing the guy and his buddies my Petty Officer Second Class patch. In port fried eggs and bacon was the usual fare for me, but at sea we ate powdered eggs and some gray, translucent strips of crap they called bacon. The chow hall had all the normal pastries and cereals too, but I wanted something hot. The Navy cooks never heard of grits, so all they fed us was lumpy, pasty oatmeal and a sickening, weak porridge called Cream-of-Wheat, which Yankees ate with sugar on it. Some of my Yankee friends in our squadron said they ate grits too, but they called it corn-meal mush and had to put sugar on it too. After washing our delectable breakfast down with coffee or powdered milk or orange Kool-Aid, we'd climb the stairs to the hangar deck. We'd check in at our shop, and over another cup of black coffee we'd find out what we had to do that day.

We always worked in pairs in case something bad happened. We couldn't wear jewelry because it could be caught on a plane's metal parts. One of our junior men lost his finger when he jumped from the cockpit ladder on the side of the plane and his wedding ring got hung on the ladder's top step. During flight ops, someone would be given the task of making a last-minute fix to a plane's computer or radar so it would be allowed to fly and not be scratched. We'd replace a part or reset a switch on the hangar deck, and the plane would be pushed and pulled to the elevator and up to the flight deck. Once we were replacing a computer Aft Processing Unit that sat between the legs of the B/N while the plane was attached to a lowboy aircraft tractor moving it slowly toward an elevator to go topside. A Chief in his yellow aircraft handler's sweatshirt, white vest and goggles ran up to me and yelled in my face, "We not gonna to stand for that, motherfucker." Too late; we had already finished, so I just smiled at him and we walked away. He could have reported us, but he had much better things to do.

After a flight's recoveries were made, we would get troubleshooters' comments and pilots' complaints about any failed systems, and we would be paired up and assigned to fix a plane by Chief Cranston or Petty Officer First Class Lash. Jets launched or recovered on an aircraft carrier were subjected to very intense starts and stops caused by being shot off the ship by the catapults or their capture at recovery by the arresting gear. The great speed at which each of those events occurred shook our four sensitive computer units like a rag doll. Quite often parts inside our heavy, metal computer boxes failed from overuse and the physical violence of launches and recoveries.

We climbed inside or stood in the wheel-wells of our planes to do our work; that's where our equipment was placed. In the cockpit we had to be careful not to ruin anything or blow ourselves out of the right seat, bombardier/navigator's seat. Our equipment was primarily for the use of the B/N, not the pilot; we only messed with the pilot's weaponry sights. On days when we did not have flight ops, we were assigned specific duties to perform on specific aircraft. This could be done on the flight deck or the hangar deck. Experienced AQBs were in charge and newer sailors helped them. The heaviest computer part we Aviation Fire Control Technicians worked with was the drum, the brains of the computer, where all our info was stored. Most guys would have to push a movable scaffold beside the plane and return it afterwards when they had to have the drum replaced. I was able to manually push it up with both arms, like doing a press with weights, to the top of the steps of the cockpit and have another AQB pull it inside the plane without the time-consuming use of the scaffold. I was usually the one told to lift up the good drums and hand down the bad drums in our planes. When it came time to eat and it was convenient, we broke for chow. We checked in with the shop beforehand, and then afterwards, to see what was next or if we were to continue with our planes. We worked until the planes were fixed or until we gave up and had to get someone else to help us. Aircraft carriers carried civilian manufacturing representatives on board at all times to solve our hardest problems dealing with their aircraft and their computer and radar systems. Sometimes a plane would be down for two or three weeks while our best men worked constantly with aircraft reps to find one damn broken wire hidden deep inside the body of a plane.

We would grab some supper when we had time toward the end of our shift. We then returned to work. Many hours were spent watching closed circuit flight operations or rerun TV shows on our tiny, black and white screen mounted on the far wall of our shop. We would work on planes in the hangar no matter the time of day or other circumstances. At the end of our shift, we would be replaced by the night shift. After telling them the status of our planes, we would get a snack and go back to our berthing compartment to read or write letters in our rack. Sometimes we might even shower before doing so. Working 12 hours on and 12 hours off meant we had to find a lot of something to do when we were not working. Reading, writing, watching the tube, standing at the elevators and the fantail looking out at the ocean; those were things we normally did when we were off. We played bridge, pinochle and poker, we smoked and drank orange Kool-Aid, and we told lies and secrets to each other. On a very rare occasion, someone would be caught trying to find love in all the wrong places. I presume they were sent home. On evenings when there were no flight ops, several of us congregated on the flight deck at the very bow of the ship and sat with our feet in the catwalk that kept us from falling into the sea and watched the water, the flying fish and "stars out the ying-yang." On the flight deck with no lights anywhere, with shipboard visibility provided by only red running lights, we could see every star in the entire Western Hemisphere in all the colors of the rainbow, or so it seemed.

18. POMPEII AND CALLEY

Toward the end of May of '71, we anchored off the coast of Naples and began preparations to hit the beach. The city looked old but exciting, and the weather was terrific. There were comparatively few tourists in Naples as opposed to the hundreds of thousands we saw later in the central and northern parts of Italy. Apparently, the Italian tourist bureaus weren't enamored with the Naples region either and barely promoted it; that gave us a chance to sightsee without being crushed by tourists loaded with money. Instead, we were crushed by the multitude of sailors on liberty in different stages of drunkenness and without any money.

My good friend Chuck from Oakland had popped into the head one morning while I was showering for liberty. He wanted to go into Naples with me and Pete, and I said, "Sure, meet us back at the AQ shop in forty-five minutes."

On the way back to my rack to get dressed, I saw Pete standing beside his. I told him Chuck wanted to go with us, and he said 'no way.'

"Why not?" I asked.

Pete looked at me and said, "Smitty, we can't pick up any girls if we have a black guy with us. It's hard enough as it is, but it'll be impossible with Chuck. I'm not wasting my time in Naples being a civil rights freedom fighter. I saw enough of that in Memphis."

"I don't agree. We won't even talk to any girls anyway."

"Look, if you want to go with Chuck, go ahead. Just remember; this isn't the Pacific. The women over here in Europe are college girls and tourists on holiday, and they don't go for black guys. Play like you're in downtown Atlanta or Birmingham. I'll be at the liberty launch in half an hour if you want to go with me."

All I could think to say was, "I didn't know you had become such an expert on European race relations and the opposite sex. Where you hear all that crap?"

Pete thought a second, smiled and sang a line from *Bennie and the Jets*, Elton John's new hit song. *You know I read it in a magazine... Bennie, Bennie, Bennie and the Jets.* He ended 'mag-i-zeeen' at the top of his voice in a screeching, high-pitched tenor and threw his hands way up high at the finish. I didn't like the message one bit, but he laughed his butt off, so at least he was in a good mood, and nobody else heard it.

Like the low-down coward I was, I told Chuck a lie when he came to the shop. I said Pete had already lined up some sightseeing trip for us, and we didn't have an extra ticket for him. Chuck looked at me and knew I was lying. He knew we weren't the kind of sailors who go on some boring tourist company sightseeing trip; I should have come up with a better lie. But I couldn't bring myself to tell him the truth: I had picked Peter Boy over him because I didn't want to jeopardize my friendship with Pete to save my friendship with Chuck. I couldn't tell him that the real reason the three of us couldn't go on liberty together was because Pete didn't want to socialize with blacks under any circumstances, although he had never complained when Chuck was with us at my apartment in base housing. Chuck didn't say anything; he didn't say kiss my butt or anything. He just turned and silently walked away, and he never asked to go on liberty with me again. I couldn't get Chuck's look of disgust out of my mind.

Pete and I went into town to have a couple of cool, refreshing beers.

We both agreed with the Tourism Bureau; there really wasn't anything to do or see in Naples unless you had some money. In reality, that was unfortunately true everywhere, at home or abroad. Pete was itching to do something fun, and I wanted us to stay out of trouble, so I encouraged him to find something for us to do or somewhere to go. He looked through a couple of travel brochures he had pulled from a rack on the street and said, "Let's go to Pompeii." We walked to the bus station and took an express to Pompeii. We arrived at the Porta Marina entrance in mid-afternoon, and I bought Jane a pretty cameo made, so they said, in the factory at the foot of the dead city.

Pete read from his brochure, "The excavated ancient city is embedded with the dust of the twenty thousand Pompeiians who died on August 24, 79 AD when nearby Mount Vesuvius erupted."

The complex was fascinating, but the one truly educational experience in Pete's measured opinion was the erect bronze twenty-inch penis which prominently jutted high above the front door of the two-thousand-year-old whorehouse to advertise its location and to welcome its customers.

"Man, that thing's scary. Looks like it's going to attack somebody; like a baseball bat or something. I hope it's not life-size; I'm a world of trouble if it is," Pete said.

As I pointed in the brochure, I said, "Of course it's not life-size...The brochure says right here that those ancient warriors were well endowed, and therefore, that thing's only half their actual size."

"Let me see that," Pete said, as he snatched the brochure from my hand. "I don't believe they'd put that in there. That can't be true."

"You know I'm just playing with you, man. You're so gullible; you'll believe anything."

Everything was going fine until Pete pulled a joint out of his pocket and lit it in public. I kept my cool and didn't say a thing. We were walking on a deserted cobble-stone street in the middle of old Pompeii, and I gave Pete a lot of room. He didn't care; he just wanted his weed. And I didn't want to be accused of smoking with him if he got stopped by the local police or the Shore Patrol, which was unlikely since we hadn't seen either in three days. I worried through the rest of the day, not about the police, but that I was becoming just another old married man, and Pete was still young and fearless. When we got back to the ship, I saw that Jane had sent me a large package. Inside were two more magazines, a Newsweek and a Time, both for April 12, and both were loaded with horrible details of the Lt. Calley My Lai stuff. Thank God I was in the Med and not on Yankee Station, or in-country Vietnam where I would have had to deal with the chaos and confusion our GIs were going through.

Almost as soon as the ship pulled out of Naples, she pulled into Civitavecchia Port in the harbor serving Rome. Civitavecchia is a mouthful, but it simply means 'ancient town,' actually 'town ancient,' which Rome certainly is. 'Civita' means town or city, and 'vecchia' is the Italian word for ancient or old. It's the same thing for the old medieval bridge crossing the Arno River in Florence; it's 'Ponte' for bridge and 'Vecchio' for old. Ponte Vecchio meaning 'old bridge,' but actually 'bridge old.' Civitavecchia's shoreline was protected by a constant stone seawall over a mile long. We anchored and took launches into the only opening in the seawall, on its west end. We traveled a good mile, got off the liberty launch, and took a forty-five-mile train ride to Rome.

My good friend Gabe had a girlfriend in Miami who was a stewardess for Pan-American Airlines and had arranged for her to meet him in Rome. She easily did that by taking a few vacation days and travelling on her employee account for which she only paid minor surcharges and fuel taxes on her free ticket. She also arranged to bring two of her girlfriends, also stewardesses, with her. The three of them were meeting Gabe in Rome, and he was two men short. Of course, he asked Pete and me to join him.

The girls were very nice and pretty; there was no hanky-panky, not that Pete didn't try. Everybody paid his or her own way. We met them at their hotel and went to the Vatican and the Sistine Chapel together. We walked around Rome and went to the Coliseum and the Senate. It was late, and we were all tired, so we headed back to the girls' hotel. We were surprised to find that Europe was quite prudish; unattached women could not have men in their hotel rooms for any reason. Peter Boy tried it and was stopped on the stairs leading to the guest rooms by a hotel employee. The girls laughed and went to their room. We sailors went back to the cheap little hotel room we had rented for the night.

The next morning, we had time to visit the Vatican again, where we ate pizza and drank wine on the Piazza San Pietro until the girls had to catch their flight back home. Jane met them in Oceana upon our return; Gabe invited them to join us at our arrival in Norfolk for one last drink. His girlfriend later let Jane and me stay at her Miami apartment for a month when we first got out of the Navy and hadn't yet decided where to live. Near the end of May, the ship anchored at Argostoli on the Greek island of Kefalonia. We had no idea why we were there; of course, we never had any idea why we were anywhere. We had no need to know, so we weren't told; it's an ancient military principle. Thank the Lord, we didn't get a chance to spend any money since we couldn't go ashore. The *Forrestal* next sailed directly to Malta to add more visitation time for the foreign dignitaries and government officials who hadn't come aboard earlier due to bad weather. The Maltese people were very friendly and looked on our visit as a major attraction. We anchored for a week to greet those lovely people and take on more fuel and supplies. Malta, an independent island nation located just south of Sicily, was primarily made up of dirt and stones. Of course, there were olive trees all over certain areas of the island just like in all Mediterranean nations. They were about the only trees we saw and were old and twisted like large scrub oaks.

Pete was getting unruly and wild, going stir-crazy from all the petty crap we had to do, and I was beginning to feel the same. In a war zone, there is very little petty stuff to do; but in a peace zone, there is very little non-petty stuff to do. To calm us both down, I agreed to join him on a thirty-minute ferry ride to the island of Gozo, the smallest of the three islands in the Malta chain. Hopefully, it would settle his nerves; he was tense and itchy and needed to go someplace quiet and peaceful. I sarcastically thought maybe some marijuana or cocaine would settle his nerves, but I kept my mouth shut; I hoped that was not his problem.

We disembarked at the Gozo Ferry Terminal at Mgarr and took a tour bus inland. Near the city of Victoria, we stood in a large open meadow surrounding a huge and ancient church, famous for its marble floors and paintings. The massive hillside it sat upon was just the tip of the giant rock formation that made up the island. The meadow was on the gentle weather-beaten slope of a small portion of the mountain, and its lush green grass was carefully manicured in an area the size of four football fields laid side by side. Some pretty wildflowers would have made the scene lovely. But, as it was, there was just a giant spread of green natural grass with a large four-hundred-year-old cathedral stuck right in the middle. No bushes, no flowers, no parking lot, and only one paved road, which may have really been packed dirt or clay.

A terrible storm unexpectedly swept in over the small mountains to the north where it had been hiding, releasing a mighty flood of cold, refreshing rain. Some folks ran to the cathedral's door, but it was locked. They hurried back to the bus, but not before they were drenched. Our driver waited for the storm to clear, but it was there to stay. We tourists were wet and bored and some complained; why hadn't he warned us about the weather? True, he should have, but in his defense, there was no Weather Channel back then; they may not have one today. The driver gave up waiting for the storm to end and started the drive back to the ferry. That was it, the whole tour; one big church in a big meadow in the rain, and we couldn't even go inside and look around. Pete complained to the driver who reached over and pulled out a brochure that said "Driving Tour Only" in tiny script on the bottom of page three.

Pete told me in front of the driver, "Now you know why I hate these damn tourist scams."

"You picked it out, bonehead. It wasn't a scam. Next time, read the whole brochure. No wonder it was so cheap; we should have known better. Anyway, let's try to have some fun the rest of the day. We'll get some donuts and hot coffee near the ferry."

"Or some beer and wine and pizza," he added.

"And nothing else," I said. He gave me a hard look; he knew what I meant, no drugs.

The storm whipped the sea outside the seawall into a frenzy of white caps. The ferryboat captain said there would be no boat service for the remainder of that day. There were no American-type motels in the Mediterranean, only hotels. There was usually not much demand for either on Gozo. Overnight tourists went there to stay in villas and private resorts. It had one large modern hotel, but not near us. Besides, neither Pete nor I had the money for a hotel room. We had no place to stay, and we couldn't get off the island. Pete talked the captain into finding a couple of pillows and two blankets for us and letting us stretch out on the long, hard wooden benches at the ferry station.

The next morning on the first accessible liberty launch, we went back to Valletta on the big island and then on to the *Forrestal*. We checked in and cleaned up. Everybody knew about the storm and that it had kept us on shore, so we weren't in any trouble. I went to see if Pete was ready to head back to town and caught up with him in the head. He was standing in a back corner like he was hiding. I asked him if he was okay, and he said yes. When he turned to me, I saw he was holding a joint between his fingers, and he said, "Here, take a hit on this. This is the best stuff I've ever had."

"You idiot, they'll kick you out if they catch you smoking that crap in here. You know, you're beginning to worry me. You're taking too many chances. What's going to happen to you?"

"Don't worry about that, buddy," he said bitterly. "No one else is around here, I checked it out; they've all gone ashore or are working. And you don't ever need to worry about me, you big married man and Sunday school teacher. I'm the same as I always was. You're the one who's changed."

"I haven't changed, Pete. I didn't start using that stuff; you did. And I am married and, therefore, I do have to worry about what I do. I can't go off and spend the night with some tramp or prostitute like you can. You don't have to worry about giving some disease to somebody you love. Of course, that's not a big problem with you because you don't have anybody to love," I said.

While we were in port, we had mail call. I stayed around to see if I had anything just because I had nothing else to do. Surprisingly, Jane had sent me a small package with a letter and a paperback book. The book, *Calley: Soldier or Killer?* by Tom Tiede, detailed a lot of Calley's life before his Army days and delved into his trial. To have something to keep me busy when we weren't flying, I began a timeline of Calley's life from Tiede's book and my four news magazines and two books from the ship's library. I wrote and rewrote the following story that my friend Gabe saw on my rack one day and asked what it was.

"This is a little story I've been writing for fun. It's about some GIs in Vietnam shooting up some people. Wanna hear it?" I asked.

"Yeah, man, I love war stories," he said.

"Well, there's something screwy about this story," I said.

I began to read out loud as he pulled up two folding chairs, "On September 7, 1967, William Laws Calley, Jr. received his commission as a Second Lieutenant in the Infantry of the United States Army, graduating 120th of the 156 officers and gentlemen in OCS Class No. 51. Reviewing Calley's military career after being commissioned makes one wonder how well the 36 men who graduated below him performed during their military service. Lieutenant Calley was assigned to the 1st Platoon, Company C, 1st Battalion, 20th Infantry Regiment, 11th Infantry Brigade and was immediately ordered to Schofield Barracks, Hawaii, the headquarters of the 11th Infantry.

"While Calley was training for combat in Hawaii at the Jungle Warfare Training Center, antiwar protesters were marching on the Pentagon. Calley and his unit didn't know about it, nor would they have cared, for they had other and more pressing things in store for them. Besides, there'd been major stateside demonstrations before, like the one two years earlier on November 27th in '65, when thousands of antiwar protesters had gathered at the Washington Mall to giddily sing protest songs, wave antiwar banners, burn draft cards, and pose for the press. That demonstration, the March on Washington for Peace in Vietnam, was organized by the Committee for a SANE Nuclear Policy and attracted 35,000 demonstrators who picketed the White House and then walked over to the Washington Monument. It was the largest public protest against the Vietnam War up to that time. The demonstrators were asked to not burn the American flag or demand an immediate withdrawal from Vietnam and to carry only committee-approved signage. The organizers were worried about alienating the American public and lowering its opinion of their antiwar movement.

"Interestingly, on that same morning of November 27th, it was announced that the entire 7th ARVN (Army of the Republic of Vietnam) Regiment of the South Vietnamese Army was killed in a battle with Viet Cong and North Vietnamese Army troops. The NVA had surrounded the 1,000 man ARVN unit at the Michelin rubber plantation in the Dau Tieng district of South Vietnam. Most of the South

Vietnamese troopers, as well as their American advisors, fought until their ammo was gone. The soldiers who surrendered or were wounded were gathered together at the site and machine-gunned to death by the Viet Cong. One has to wonder who would have cared most about that information: American citizens, American servicemen or American antiwar demonstrators.

"During the next two years, many protests, big and small, took place all over America. In 1967, on the morning of October 21, at the time Calley was rising in preparation for another training day in Hawaiian jungles, organizers of the march on the Pentagon announced their schedule for one of the largest. The National Mobilization Committee to End the War in Vietnam informed their protesters they would be marching as a group to the Lincoln Memorial, then across Memorial Bridge, all the way to the Pentagon. Abbie Hoffman, the founder of the Yippies, the Youth International Party, told the exuberant crowd they would pour red dye all over the Pentagon and would burn its cherry trees. He said they would "piss on the Pentagon" walls and "raise a flag of nothingness over it and a mighty cheer will echo throughout the land." The capital was packed with about 100,000 or so primarily young white kids who were exhilarated by participating in the antiwar happening. Some had come to express their convictions and some had come to party. Around 3:00 p.m. EDT that Saturday the 27th, many protesters began leaving the rest of the gang at the Lincoln Memorial, but almost 50,000 of them continued trudging on to the Pentagon, arriving around 5:40. The rally soon turned to riot. By 6:00, some of the demonstrators tried to storm the Pentagon building. Some 30 or so pushed through the line of U. S. Marshals and military police at the entrance but were restrained and forced outside the building by heavily armed soldiers who formed the second line of defense. The riot lasted all night, and by 7:00 a.m. Sunday morning, only 200 protesters remained.

"A total of 682 people were arrested and 47 demonstrators, soldiers and U. S. Marshals were injured during the protest. The Deputy Marshals, the federal government's civil enforcement authority, made all the arrests so that the world could see our concept of civilian supremacy but also our commitment to civilian control. When arrested for various degrees of civil disobedience, many demonstrators collapsed to the floor, making the Deputies drag them outside to prison vans where they were pushed and shoved aboard against their wills. The Deputies handled the protesters with increasingly rough treatment throughout the night; they had worked over 14 hours with no relief and were exhausted and disgusted. The Pentagon protest came one month before the November 20th announcement by the government that the number of Americans killed in Vietnam had reached 15,000.

"After almost five months of training in Hawaii learning to command troops and to survive in Vietnam, Calley with his company was flown to Da Nang and arrived December 1, 1967. Heading for Vietnam's Quang Ngai province, the men of Charlie Company, made up of three rifle, one weapons and one headquarters platoons, were not overly concerned. Not yet. Calley was the Lieutenant of the 1st Platoon of Charlie Company, First Battalion, 20th Infantry Regiment, 11th Infantry Brigade, 23rd Infantry Division, United States Army. Captain Ernest 'Mad Dog' Medina, Charlie Company's leader, was known for his quick temper when his expectations were not met. Medina looked down on Calley, belittled him and called him 'Sweetheart' at times. Calley looked up to Medina and would do anything to please him; he felt insecure and inferior and wanted Medina's approval and respect. He would follow any of his orders.

"Instructors from the 4th Infantry Division Noncommissioned Officer Academy oriented the men of Charlie Company on December 3rd on basic patrol techniques, on how to call in fire support and how to handle prisoners. They were also taught how to distinguish Viet

Cong from civilians. Charlie Company was airlifted to Landing Zone Bronco in Quang Ngai Province, about eighty miles south of Da Nang, where soldiers of the 2nd ARVN Division taught them how to recognize booby traps and mines and how to assault and secure a village. A few days later, the men moved to Landing Zone Carrington, set up the 11th Brigade fire base there and practiced patrol and search and destroy missions. With no enemy contact and lots of free time, the men went to the beaches and the bars, drinking beer and getting to know the boom-boom girls.

"Charlie Company was moved to Landing Zone Dottie, a well-fortified complex of sturdy bunkers. During the Tet Offensive, which began January 3, 1968, the men watched the fighting for Quang Ngai City in comparative quietness at their remote fire base north of the city. Although several casualties were suffered on patrols, mostly from booby traps and mines, the action swirled around them like a swift river flowing around an island in its midst. On patrol on February the 12th, Calley led his men to a place where they had incurred VC small-arms fire the day before. Sniper-fire was again encountered and Calley changed their route, stupidly letting his men walk along the top of an earthen dike exposing them to enemy fire. Suddenly, from nowhere, a bullet tore through the kidney of SP4 William Weber, Calley's radio telephone operator, who fell and bled to death in minutes. Calley never forgave himself for his inexcusable lack of judgment and the death of Weber. To justify Weber's death, Calley filed a false action report that night, saying that his platoon had killed six VC in the firefight in which Weber was killed.

"As one of the three companies making up Task Force Barker, Charlie Company had been sent to LZ Dottie in an area commonly known as the stronghold of the infamous Viet Cong 48th Force Battalion. The 48th retreated eastward from their Tet Offensive positions toward Quang Ngai's coast on the South China Sea in early February of '68. Charlie Company increased its patrols in hopes of making contact with the Viet Cong but to no avail. A few men of Charlie Company began to break down from the tension caused by knowing the 48th was in the area and from the constant screaming of artillery fire. Captain Medina held Calley personally at fault for not holding them together.

"On February 13, Task Force Barker's three platoons finally engaged the enemy. In a three-day operation designed to trap the 48th VC Battalion, Alpha, Bravo and Charlie Companies killed 80 Viet Cong while suffering three Americans killed and 15 wounded. No weapons were recovered. On the 23rd, Alpha and Bravo Companies made contact with the 48th and killed 78 VC while incurring three American deaths and 28 wounded. But the 48th remained intact, active, nearby and deadly dangerous.

"On the 25th of February, Charlie Company's 1st and 2nd Platoons blundered into a minefield, killing three men and injuring 16 more. The first explosion blew men and equipment into the air and caused other troops to panic and run, resulting in more explosions. Capt. Medina was awarded the Silver Star for his cool head and quick actions to get medical attention to his men and to summon mine sweepers to get his men out. The grunts were angered and traumatized at the loss of their friends and were frustrated that they had no way to release their tension and heartache and no one on whom to take revenge. They began to talk about how stupid Calley was; some said he was terrible at reading a compass or a map. Some even talked of fragging him, if only they could catch him alone. The strain showed; the GIs began to treat the Vietnamese more harshly, even the children. Most of the men believed the Vietnamese knew of the minefield and had not warned them. In their minds, they began to believe the Vietnamese around them were as much Viet Cong as the VC who had placed those mines."

"That's it," I said. "That's all I have right now."

"Man, there's a lot of crap in there," said Gabe. "How long you been writing it?"

"Maybe a month, I guess." I smiled and stuck it under my fart sack, safe and secure.

"Somebody's in a world of trouble," Gabe whistled. "Let me know what happens."

19. CORFU AND MY LAI

Pete and I didn't talk much for a while after I caught him smoking weed in the head and we'd had our blow up. But in early June when the *Forrestal* anchored at Corfu, an island due west of the Greek and Albanian border, Pete wanted to go sightseeing again. So I went ashore with him just to keep him straight. Neither of us had more than a few dollars to spend on anything other than beer and food, except he did have a little marijuana money, he told me later. I hoped he wasn't doing anything else.

"I thought you didn't want to go on any more tourist scams. What in the hell are we going to do on Corfu? The place isn't as big as Georgia's campus."

"Hey, I checked it out...good. It doesn't cost anything to get there; the liberty launch will take us to this beach I found. We can buy some cheap sandwiches and a few bottles of beer when we get there. It'll be fun."

"Yeah, I'm sure it will," I muttered sarcastically. "But what will we do? Go swimming? Or boating? Or fishing?"

"Yeah," he said. "All of the above. Maybe we'll just drink beer under some palm trees on the beach."

"Okay," I said. "There're no palm trees in the Mediterranean, except a little dwarf palm or some goofy European fanlike palms. They're not like ours back home, and they don't grow on the beach."

"Fine, we'll just drink beer. That's all we do anyway. Every place we go, all we ever do is drink beer. Eat, drink beer, and poop. Eat, drink beer, and poop," he said.

"So who's complaining?"

"Not me."

"By the way, you left out 'work.' It's eat, drink beer, poop and work."

"Oh, yeah."

We finally planned on just going swimming; that's what the brochures said to do, so we wore our bathing suits under our civilian clothes. We had been allowed to wear civilian clothes since February, but the clothes had to be clean and non-offensive to the locals or other tourists. Sailors were prohibited from leaving the ship if their clothes weren't presentable or had a bunch of hippie crap on them like peace symbols, Hollywood actors, Black panthers, tie-dyed shirts, beaded vests, anti-war slogans, peasant blouses, Confederate flags, you name it. The officer on the gangway, or quartermaster deck, would really look sailors over whenever they came aboard or were leaving the ship. We always had to salute the American flag, it's called the National Ensign in the Navy, and ask permission from the officer of the deck to enter or leave a warship. We were closely observed when doing both. Like most anything, whenever a new rule took effect, the first few weeks were the strictest. One evening I came aboard without any shoes, but that was in the Philippines a year earlier. The small sailboat Pete and I were on in Subic Bay had capsized, and my navy dress shoes which I had taken off for comfort, went into the sea, never to be seen again. A Filipino fisherman straightened us up and pulled us back, half full of sea water, to the base's recreation area.

We ate a delicious local peasant sandwich on Corfu, drank a couple of cold Italian beers and walked to a secluded beach nestled between some cliffs. There were a few people lying in the sun and swimming in the light green water. We just stood there awhile because there were no bath houses where we could shed our clothes. As we looked around, we saw a couple walk down to the beach with what looked like two five-year-olds. They were all dressed in street clothes. The man took each child by the arm and held him up while the woman undressed the kid and put his bathing shorts on.

The man then held a large beach towel around his wife at shoulder height, and she completely undressed inside the towel, right there in front of us. It was no big deal; we couldn't see anything, but we had been without female companionship so long, we took notice of every small move she made. She bent down and grabbed her bathing suit and put it on. I hope our mouths weren't wide open. She took the towel from the man and held it around his chest. He undressed and put his swimsuit on, too. The lady picked up their clothes, folded them neatly, laid them on their shoes, and then all four of them ran into the water.

Well, after seeing that, Pete and I played follow the leader and opened the beach towel we had bought in town to lie on. I held it around him while he took his street clothes off, and he did the same for me. It was exciting to enjoy the different cultures and styles of living practiced by other people of other countries.

The *Forrestal* pulled out of Corfu on the 11th of June and sailed farther out into the Ionian Sea. She wandered around for a couple of days and then headed to Crete, the largest Greek island, directly south of Athens. I had been working on my Lt. Calley at My Lai story and had finished for the day. As I was putting it under my rack, Gabe came up and saw me.

"Hey, man, whatcha been doing? You ever get that story finished?" he greeted me.

"My man, what's up? Yeah, I've written a little more. Wanna hear it?" I asked.

"Sure, I liked it."

I picked the story up and started reading. "In early March, Army intelligence reported the Viet Cong 48th Battalion was almost certainly in Task Force Barker's area near the villages of My Lai. Lieutenant Barker and his staff believed the Viet Cong battalion could be wiped out with a surprise attack while they were resting and replenishing. On the 14th of March, while on patrol, Sergeant George Cox was killed by a booby trap, and eighteen other soldiers were injured, two seriously. As they returned to camp from the patrol, several men shot and killed a female Vietnamese civilian working in a field, the first documented incident in Vietnam of a violation of the military code for killing a human without justification or cause.

"Calley had just returned from R & R at a nearby seaside resort and helped unload the company's gear from a chopper. He held back tears as he later said, 'The thing that really hit me hard were just the heavy boots. There must have been six boots there with the feet still in them. Brains all over the place. And everything was saturated with blood. Just rifles blown in half. I believe there was one arm on it. And a piece of a man's face.' At Sgt. Cox's packed memorial service the next morning, Medina told his men to wait; they would soon have the chance to exact revenge for his and their other friends' earlier deaths. According to 2nd Platoon PFC Dennis Bunning, Medina said, 'We're going to get even with them for all the losses we've had. We're going in there; we're killing everything that's alive. We're throwing the bodies down the wells, we're burning the villages, and we're wiping them off of the map.' That was good news for the men of Charlie Company; they had suffered 28 casualties, with five dead and others maimed for life, and had not once met the enemy face to face.

"At a midday briefing on March 15th, Medina and other company commanders were erroneously told that the small group of hamlets called My Lai was presently providing a safe haven for the VC 48th Battalion. There were few Viet Cong soldiers in My Lai at the time. The time had come to attack, and the GIs were told they would soon meet the enemy and were encouraged to be more aggressive, very aggressive. They were told that any neutral Vietnamese civilians in the village would have already left for the markets seven miles away in Quang Ngai City, and only the Viet Cong and their civilian sympathizers would be still be there. Later that day, reconnaissance flights were made over My Lai, one of the four hamlets in the village of Son My, to identify landing zones.

"At 5:00, Captain Medina personally gave Charlie Company its orders for the following day, purposely exaggerating the number of expected enemy soldiers hiding outside of Son My and telling the men they'd be outnumbered two to one. Medina said they would attack Son My and destroy all the crops, kill all the livestock and burn all the dwellings. Did Captain Earl Michles give his Bravo Company the same orders for the same mission? Michles said 'sharks,' Hiller OH-23 Raven helicopter scout teams, would be overhead surveying the area. One of those shark pilots was Warrant Officer Hugh Thompson of the 123rd Aviation Battalion of the 23rd Infantry Division who would become a major player in the next day's activities. Thompson was a 25-year-old U.S. Navy veteran who had served honorably for three years and had gone back home to Stone Mountain, Georgia. When Vietnam began flaming out of control, he felt it was his duty to fight and joined the Army to fly helicopters. An actual Boy Scout had become an Army scout.

"At 5:30 a.m. on March 16, 1968, a day which would become as infamous as the Japanese surprise attack on Pearl Harbor, Charlie Company rolled out of their bunks. Up until that moment and for another two and a half hours, there had been nothing unusual or unique about the men who made up Charlie Company. Their personnel records, their field training, their combat performance, everything, were as normal as any other unit in the Army, but the beginning of the company's end began at 7:22 that morning. As the nine helicopters of the first element of the My Lai mission carried Calley's 1st Platoon and most of Lieutenant Stephen Brooks' 2nd Platoon from LZ Dottie, Delta Company of the 11th Artillery Battalion fired into Son My to clear a landing zone. That presumably necessary action unfortunately caused the villagers to run back into the village instead of leaving for the markets in Quang Ngai City. The first element landed in a couple of paddy fields outside the village with a few unconfirmed and questionable reports of enemy fire. The second helicopter lift took off for My Lai at 7:38 a.m. with the rest of the 2nd Platoon, Lieutenant Jeffrey LaCross' 3rd Platoon and, as fate would have it, an army photographer assigned to Charlie Company for the day. Calley moved his men along the western edge of the village into a defensive position securing the landing zone for the lift, and Brooks led his men to the northwest edge. Several villagers ran from their hiding spots and were shot. While flying in their scouting position, Warrant Officer Thompson's door gunners shot at several armed pajama-clad Viet Cong running down an earthen path, but the crew was unable to see if they hit any. At 7:47, the second element of helicopters and troops landed and reported receiving fire. The landing zone was declared "hot." The scouting sharks circling the village killed four armed Viet Cong on trails leaving the village. How many VC were there?

"American troops began moving through the village at 7:50 a.m., only 28 minutes after their lift off from LZ Dottie to begin the mission. The two platoons from the first helicopter element approached My Lai with Sergeant David Mitchell leading the first squad, quickly followed by Calley with a squad of 24 men and then two other squads. The GIs shot and bayoneted many of the fleeing Vietnamese, not knowing if they were civilians or VC, and threw hand grenades into houses and bunkers. As ordered, they destroyed all the livestock and crops they could find. They began rounding up people, approximately 40 or 50 civilians, mostly old men, women and children, and pushed them to a dirt road south of the village. Another 70 Vietnamese were moved to the east of the village. Soldiers of the two platoons suddenly began firing on the civilians for no good reason and with no provocation. Many of the Vietnamese men were stabbed with bayonets or shot in the head. As they prayed in front of a temple, over a dozen women and children were shot and killed by passing soldiers.

"Captain Medina called operations and reported 15 VC had been killed. Unfortunately, 400 or 500 civilians fleeing to Quang Ngai City happened to be intercepted by one of Lieutenant LaCross' squadrons and were fired upon, killing 10 to 15 of them. Calley's platoon entered Son My village from the south and joined the slaughter. A wounded old man was shot in the head by one of Calley's men, who passed it off as an act of mercy. A woman carrying her baby and holding her toddler's hand was shot and killed. An elderly woman ran down a path with an unexploded M79 grenade lodged in her stomach. In an insane act of depravity, one GI forced a 20-year-old woman to perform oral sex on him while he held a pistol against her wide-eyed, hysterical 4-year-old's head. A day of infamy, indeed. Captain Medina reported to Lt. Col. Barker at 8:30 a.m. that 84 of the enemy had been killed. Barker passed it on to the tactical operations center.

"PFC Paul Meadlo and Pvt. Dennis Conti stood guard over the group of 40 or 50 civilians on the road south of the village. Calley came up and told them, 'You know what to do with 'em.' Meadlo thought that meant to watch them. Calley returned ten minutes later and screamed at the men because the villagers were still alive. Calley, according to Meadlo, said he wanted them dead; all of them dead. Calley stepped back a few feet and began shooting randomly into the crowd. The men said Calley ordered them to join him and kill them all. Meadlo stood 15 feet away from the civilians and poured a minimum of four clips of 17 rounds each into them. Meadlo and Conti then rounded up another 50 Vietnamese and moved them into a ditch. Lt. Brooks'

2nd Platoon moved throughout the village, shooting 50 to 100 fleeing civilians without exception and committing two rapes.

"The 3rd Platoon entered Son My at 8:45, destroying and burning everything, and shot and killed a group of 10 or 12 women and children. Scouting shark Warrant Officer Thompson headed to LZ Dottie to refuel and flew over a large area of Vietnamese bodies, mostly old people and children. Confused and not understanding what was actually happening, the crew thought artillery had killed them all. They saw a wounded woman on the ground and marked her as a noncombatant with green smoke and radioed soldiers on the ground to help them, but it was wasted breath. By the time the message had been passed along and garbled like the children's game, it told Lt. Col. Barker that eight or nine 'dinks' with web gear had been wounded. So Barker sent Capt. Medina to recover their equipment. Lt. Calley reached the drainage ditch holding the civilians at 9:00 a.m. and ordered them to be killed. Ten minutes later, all 75 to 150 unarmed Vietnamese were shot and killed by 1st Platoon members. Some of the men from Charlie Company interrogated an old man who told them that 30 or 40 VC had spent the night in My Lai but had left before the assault began.

"Upon their return from refueling, Larry Colburn, Thompson's helicopter gunner, watched as Capt. Medina walked over to the injured woman they had marked with smoke and killed her. That's when Thompson and his crew really knew what was happening. 'It was our guys doing the killing.' Around 9:20, Thompson's crew witnessed a sergeant shooting people in a ditch. Thompson dismounted and approached Calley who was standing nearby and asked what was going on. Thompson said there were too many bodies there and something was wrong. Calley said nothing was wrong, he was just following orders. But these are just women and children, Thompson said. Calley said he was following his orders and it was none of Thompson's business. He told Thompson he'd better get back in his chopper and get out of there. Thompson told Calley he hadn't heard the last of him and flew eastward.

"At 9:40, Thompson's crew saw a couple of soldiers approach another wounded girl on the ground; Captain Medina pushed her with his foot to see if she were alive and then shot her in the head. Thompson saw dozens of bodies in an irrigation ditch, some still crawling and twitching, and landed to encourage a sergeant to help the survivors. The soldier flippantly told Thompson that the only way he could help those people was by putting them out of their misery, saying the only good 'gook' was a dead 'gook.' Thompson then flew to the northeast corner of the village where 12 to 15 women and children were trying to outrun several of Charlie Company's 2nd Platoon troopers to a homemade bomb shelter. Thompson decided to do something admirable but possibly crazy. Without authority and against his training and the military concept of friend versus enemy, Thompson dropped his helicopter between the grunts and the civilians and radioed that he needed backup. Thompson's two crew members actually held their weapons on the GIs of Charlie Company, and Thompson told them to fire on our soldiers if they started firing on the civilians. Larry Colburn yelled back that they had him covered. Thompson then jumped out of his helicopter and confronted Lt. Brooks who just looked on as Thompson coaxed the people out of the ditch and onto the chopper. Thankfully, the men of Charlie Company did not fire their weapons. At 9:50, Thompson sent out a call for other gunships to come and help rescue more civilians. Medevac pilots Don Millians and Brian Livingston responded to their friend's call and lowering their UH-1 Hueys near the ditch flew nine or 10 civilians to safety.

"A couple of minutes later, Capt. Medina decided he had seen enough destruction and death and radioed the 2nd Platoon to cease fire and stop the killing. That order was never given to the 1st or 3rd Platoons, who continued killing for another hour. Around 10:00, Lt.

LaCross' 3rd Platoon surrounded 10 women and children, and a few GIs started sexually abusing a 15-year-old girl until the army photographer, Sergeant Ron Haeberle, began taking pictures. After Haeberle left, the men killed the entire group. While Medina and LaCross talked, some of Medina's command members walked off and killed several wounded Vietnamese. The company's first and only casualty occurred at 10:20, when Private Herbert Carter reportedly shot himself in the foot by accident while clearing his pistol; other reports said he purposely shot himself to avoid participating in the massacre. He was med-evacuated by Barker's command helicopter to LZ Dottie. At 11:00 a.m., Medina walked through My Lai assessing the damage and then casually stopped and ordered a lunch break for the entire company. His platoon leaders told him 90 enemy soldiers were killed, which he reported to Task Force Barker, but a more accurate count was 350 to 500 civilian casualties.

"Warrant Officer Thompson, back at LZ Dottie, told his section leader, Capt. Barry Lloyd, what had happened. Thompson then shared his concerns with Major Fred Watke in the aviation section's van and Watke told Lt. Col. Barker. Major Charles Calhoun was sent to find out what was going on and to stop it if it was wrong. Still in the field at 1:30 p.m., Charlie Company picked up ten young Vietnamese men for interrogation as they passed through another Son My hamlet. At 3:30, Col. Henderson got another report of more civilian casualties and ordered a company to go to Son My for an exact count, but Major General Sam Koster, the Americal Division (23rd Infantry Division) commander, countermanded the order, saying no further examination is necessary. Charlie Company arrived at its night defensive position at 4:00 p.m. and during interrogation, two of the ten VC suspects were killed by an intelligence officer, Capt. Eugene Kotouc. That night, Major Watke reported the day's accusations to Lt. Col. John Holladay.

"Thank God, March 17th arrived. Bravo and Charlie Companies

were ordered to search for the Viet Cong 48th Battalion. Charlie Company continued burning buildings and assaulting civilians. At 9:30 a.m., as the troops moved down the peninsula along the Song Tra Khuc River, Calley ordered his men to press on, although nothing was done to detect land mines. While walking point, PFC Meadlo stepped on an enemy mine which blew his entire foot off. As Meadlo was evacuated, he turned and shouted to Calley, 'God got me; he'll get you too, for what we did.' Some officers familiar with the story began to suggest a more thorough investigation of the alleged atrocities was needed. Charlie Company burned the hamlets of My Khe 3 and My Khe 1 and killed two Viet Cong soldiers.

"On the 18th of March, Thompson reported to Col. Henderson and discussed the unnecessary killing of unarmed civilians. Afterwards, due to Henderson's lack of concern, Thompson ripped off his pilot's wings and threw them on the floor. Brig. Gen. Young, Col. Henderson, Lt. Col Barker, Major Watke and Lt. Col. Holladay held a crisis meeting at LZ Dottie, and Henderson was instructed to conduct an investigation. His meaningless, irresponsible and slanderous report, completed a month later, concluded that only 20 civilians had been killed and Thompson's claims were false. That afternoon of the 18th, Charlie Company was airlifted to LZ Dottie and the My Lai mission was finished. Henderson interrogated Capt. Medina who subsequently told his men to remain silent about the mission.

"On March 22, 1968, a village chief reported that 570 civilians were killed in the Son Tinh District and 90% of the Son My village was destroyed. Lt. Col. Barker had a slightly different take on it. He reported that 'this operation was well-planned, well executed and successful. Friendly casualties were light and the enemy suffered heavily. The infantry unit on the ground and helicopters were able to assist civilians in leaving the area.' If those old men, women and children were indeed armed Viet Cong combatants, then they were the enemy, and Lt. Col.

Barker was correct in stating they suffered heavily. If they were, in fact, just innocent old men, women and children, then God have mercy on our souls. Of the 570 civilians killed by Task Force Barker, 210 were 12 years old or younger. Of those 210, nearly 50 were three or under. GIs had been called 'baby-killers' since '66, but the My Lai massacre cemented an image of an Army of drug-addicted American soldiers wantonly killing Vietnamese babies in the minds of the American public. In fact, it was a rare occasion.

"Charlie Company became ancient history on April 8th; Task Force Barker was declared a success and its companies were disbanded. On the 11th, Son My chiefs reported to Viet Cong officials that 400 My Lai and 90 My Khe civilians had been killed on March 16th by U.S. Army forces. On April 20th, Private Butch Gruver ran into a GI who had trained with Charlie Company in Hawaii and asked him if he had heard of the massacre. Ron Ridenhour, who had served with another unit as a door gunner, said no, he hadn't heard a thing. Gruver told him 300 to 400 civilians had been shot and killed in cold blood, and Ridenhour wanted to know more.

"Col. Henderson's report on Warrant Officer Thompson's allegations was completed on April 24th and concluded that 20 civilians had been killed accidently and that Thompson was a liar. Major General Samuel Koster, slated for Lieutenant General, found that report inadequate and ordered Col. Henderson to hold a formal inquiry, but Lt. Col. Barker took charge of the inquiry, even though his own task force was being investigated. Barker's investigation miraculously concurred with Henderson's that Thompson was a liar.

"During April and May of '68, Warrant Officer Thompson ran into a string of bad luck. He had been criticized, condemned and ostracized by many in the military community, up and down the ladder, for speaking out about the events that occurred around Son My's hamlets. He was increasingly put into dangerous situations, but those were just coincidences, some said. Shot down four times previously, Thompson was finally sidelined by his last mission, a flight from Da Nang to an airbase at Chu Lai. During that flight, he was shot down a fifth time and broke his back. A month after Thompson's crash, in June of '68, Task Force Commander Colonel Barker and Captain Michles of Bravo Company were killed in an aircraft accident.

"Ron Ridenhour, the door gunner, began to seek out Charlie Company members as he investigated the massacre first-hand while still in Nam. Sgt. Larry LaCroix told him at a June meeting that Calley fired into civilians with a machine gun. Pvt. Michael Bernhardt told him over lunch in November of '68 that he had refused to take part in the massacre. Ridenhour was discharged soon after Bernhardt's interview and returned to his home in Phoenix, becoming a journalist to continue his investigation. On March 29, 1969, a year after My Lai, Ridenhour had enough concrete information to write letters to 30 major politicians in Washington. Mo Udall's office called for an investigation and Ridenhour's letter was forwarded to General William Westmoreland, the Army's Chief of Staff. On April 23rd of '69, the Office of the Inspector General began a full inquiry."

"I've still got a lot of stuff to read," I said.

"That's cool, man, keep going," Gabe said. "I've got to see the Chief. See you."

20. CRETE AND COLLISION AT SEA

When we reached Crete, we anchored near the naval base at Souda Bay, and liberty was restricted for the ship's crew, but for some strange reason which is still unknown to me today, some of the guys from our squadron and one other were allowed to go ashore. The ship's crew threw netting down the side of the ship, and Pete and I climbed down into the waiting LST taking us to shore. The twenty of us who went on that little adventure stood on the ship's slightly rusty old hull, leaning with one hand against its side for balance. The old relic, an authentic U. S. naval vessel left on Crete after World War II, bobbed up and down the entire forty minutes it took to make the journey to shore. The LST landed just like it was D-Day on Normandy Beach but without the Nazis up on the cliffs trying to kill us with machine gun fire. It rammed itself up on the beach and dropped its front end onto the sand with a great squeal of rusty hinges. We piled out and slowly stumbled up a hill to find a nice cool pavilion perched on top with an old dilapidated shed near it. A dozen older couples, almost ancient in my book, were up there sitting under an arbor in lawn chairs, enjoying the breeze out of the sun.

Talk about absolutely nothing to do. Pete said, "This scene reminds me of going to a family reunion when I was eight or nine, and no one except me was under about forty or fifty. My mother had gotten me dressed up so she could show me off, and I looked like a fool. All the old people chunked me under my chin and said stupid things. None of my cousins had come, so I was all alone. The old farts sat around in lawn chairs just like this and looked at the birds flying by. I was as miserable as I am now. This is ridiculous. When can we get out of here?"

Unfortunately, Pete was going to get more miserable as time wore on. We stayed on that pavilion or walked around the beach for a couple of hours. Don't ask me why; maybe we were being silently punished for some unnamed offense. Clouds were moving in, and the temperature was dropping. A cold front was moving through. The old people got up, walked downhill to their little boat and left. But we still sat there. It began to rain, and I mean it began to pour. The wind howled, and the rain came down in sheets. It hadn't let up in over an hour, so the petty officer in charge of our excursion walked over to call someone on the phone from an old Cretan jeep parked by the door. The jeep's phone was dead, and the shed was padlocked, but he found a dirty military landline phone mounted on the wall, and it worked. He requested some Army trucks from the Cretan National Guard to come rescue us and take us to a place to get shelter from the storm. When the Guardsmen arrived, we jumped in, and the trucks took off. The old people had disappeared long ago through the olive trees and into their boats. They may have been caretakers who lived near there or just retirees enjoying the day. The open-air pavilion was left all alone.

We were taken to an abandoned Army barracks located about twenty miles from where we had been. We jumped out and filed into the small building. It was empty except for six bunks made of iron and some kind of wire a little heavier than chicken wire. That was nice, except there were twenty of us and only six bunks. The sun had gone down, and the rain and darkness brought a sudden coolness. The petty officer tried to fire up an old furnace, but there was no oil. And to cap it off, there were no linens, blankets, rugs, towels, or anything else to protect us from the cold creeping up on us.

When we left the ship that morning we were dressed for a pleasant, warm day at the beach and nothing else. We had planned to be back on the ship for supper, filled with gratitude, just a little, for whatever we were having to eat. That was a far cry from what really happened. The howling storm got stronger, louder, and colder. The guys who had carried us to the barracks had dumped us out and hurried off, but they still had to be somewhere on the island. Why didn't they bring us some blankets and some food? Maybe they lived on the other side of the island and didn't want to spend their entire night worrying with us. Hell, we were grown men, sailors in the U.S. Navy; couldn't we survive one cold, wet night without the comforts of home?

Most of the night I tried to find some sleep by lying on the tile floor of the bathroom. Because there was a light in there, I thought it would give me some heat, but it didn't. I was freezing to death surrounded by three toilets that were occasionally being used by sailors. Fortunately, we had not been furnished with any alcoholic beverages, or any other kind of drink for that matter, the entire time we were there. The good news was the sinks worked and warm orange water flowed in abundance from the faucets. It was hard to differentiate the smell coming out of the faucets from the smell coming out of the toilets. Maybe the guy on board the *Forrestal* who made our arrangements talked about the day's plan of action to someone who couldn't speak a word of English.

Anyway, we were miserable. I got up finally and went to find a place where I could get some heat. I had given up on getting any sleep; I now just wanted warmth. I found Pete lying on the bare metal coils of some box springs which sat crooked on a bunk frame with Dickie Clayborn huddled next to him like lovers. I liked the idea and pushed in between them with my feet to their faces and their feet in mine. There was no other way the three of us could keep from rolling off the bunk. I never went to sleep, but I was a little warmer. Not warm, but warmer. We were three sailors lying together on a metal cot packed in like sardines in the freezing cold, just like the rest of the guys.

The next morning found us tired, dirty, and sleepy. Our bodies were aching, and we were hungry. When a smallish tour bus finally arrived, we were loaded on it without a word being said by anyone. All we wanted was to get back to the ship and shower and shave, and the bus driver could speak no English. The ride was nice; the trees and bushes were green and lush, and the flowers were beautiful. However, no one cared, except me; the bus was quite warm, and my fellow sailors were all asleep the moment they plopped onto the padded seats. We crossed over the narrow bridge straddling the deepest valley on the island, and the bus's engine whined as it labored pulling us up the span. I looked outside as we hit the surface of the bridge and noticed the retaining wall and support railings on the opposite side were torn away and discolored. It looked as if there had been an accident at the point where the bridge lifted off the ground. I tried to see what had happened because it looked so new. On my left side, the bridge itself kept me from seeing into the gorge below. On the right, my view down into the river was obstructed by the guardrail that ran the length of the bridge. I couldn't see the lights and ambulances at the bottom of the ravine. It was funny; the guardrail looked like it was made of a cheap thin sheet-metal or aluminum. I thought that rinky-dink rail wouldn't keep something as small as a motorcycle on that bridge if it were heading off the edge.

The twenty of us bounced around from side to side in the small bus, but we finally reached our destination, where we were surprised. The Navy hadn't even sent a liberty launch to pick us up after our ordeal. We found ourselves right back on the same beach from the morning before with the same LST waiting for us to get aboard. The old World

War II relic bounced us up and down in fairly heavy seas until we reached the side of the *Forrestal*, but no one got sick. We hadn't eaten or had anything to drink for the last twenty-four hours, so we weren't nauseous, and the wimps had finally gotten their sea legs. Inside the LST, we had to move from side to side to stay out of the sloshing rain accumulated in the bottom of the boat from the night before. We used the side netting to climb aboard just as we had when we departed. The trip was a great adventure going back in time to the old Navy days, but we were tired and miserable. The ship raised its anchor, and we got underway. Either she had been waiting for us, or had come back to get us.

Sailors never heard anything official on an aircraft carrier unless they had that need to know. Obviously, that should be true of all secret information and troop and ship movements, but on this ship, it seemed to apply to everything. Scuttlebutt was our only source of "official" Navy information. Scuttlebutt, rumors, secondhand information, innuendos, lies, and gossip; that's how we heard what had happened to us on Crete.

"Hey, guys, welcome aboard. Tough night, huh? Sorry to hear about your buddies," the Chief said.

"Our buddies? What buddies?" I said.

"The guys who went to get you last night. The Cretans killed in the wreck."

The three of us from our AQ shop had gone just to check in with the Chief, and we just stared at him. "Chief, what's that? We haven't heard about a wreck or any dead guys."

Chief Cranston, surprised at my comment, shook his head in disbelief and said, "The truck going to pick you guys up at the barracks last night hydroplaned and crashed into the guardrail on some bridge. The rail gave way, and the truck fell into the river. All three soldiers died in the fall. No, two were killed in the fall, and the third man drowned when he couldn't get out of the cab. I'm sorry, I thought you knew."

"That's what happened. Those guys were coming to bring us some food, and we cussed them all night long. We may be a bunch of dummies, but somebody should have told us before now," Pete said.

The *Forrestal* left Souda Bay and began flight operations around the western Mediterranean. Things were quiet, and one beautiful Sunday at midday while in the middle of the sea, the Captain just stopped her to relax. The ship's crew set up barbeque grills on the flight deck and began cooking steaks for any and all takers. Pete and I went above deck and helped ourselves. Steaks, baked potatoes, and salads. Better than Bonanza or Western Sizzlin'.

Two days later, back at work, we were putting the final touches on the day's flight ops. We had finished recovering our planes, and the crews had moved and re-spotted them. Everything had been cleaned up, and the flight deck was clear. The Captain got on the horn and said, "Men, if you want to see a sight, come on up to the flight deck. We have intercepted a Russian Bear heading our way. It's nothing to be alarmed about; it happens almost every cruise. The Russkies just want to keep an eye on us and let us know that they're still here. Come on up; you'll enjoy it."

The Russian Bear was the Tupolev Tu-95 four-engine turboprop-powered strategic bomber used by the Soviet Union for bombing and reconnaissance missions. She was enormous with a wingspan of 164 feet, which is 55 yards. That is half a football field from wingtip to wingtip, and she could cruise at 575 mph.

A couple of hundred of us went on deck. The ship had the Bear on radar, of course; she had been picked up 250 miles away. We were told over the loudspeaker that she would be approaching in five or six minutes from the stern. This was just a little game that we Americans and the Russians played almost any time we had a carrier in the Med. Some of the guys on the fringe of our little shipboard sightseeing group shouted when they saw her. And then seconds later there she was, flying right over the *Forrestal*'s flight deck, taking pictures a mile a minute from her underbelly cameras, as she leisurely cruised over us. She got some tremendous pictures during her flight, but unfortunately for her, not of our flight deck; instead, she took a thousand pictures of the bottom of the F4C Phantom flying upside down directly beneath her. We watched as those two planes flew two or three thousand feet above us, the Bear on top and the Phantom underneath, only yards separating them, like a momma bear and her cub. Our jet peeled away after they passed the ship but remained with the Bear until she was far away.

About this time, we received some guests on board. We had opened the ship up for guests during the entire deployment, but this was different; these folks came from Hollywood to film a major motion picture named *Carrier*. This was the right place at the right time because we weren't doing anything. Afterwards, the movie didn't do anything, either. Apparently, it was never made or released under *Carrier* or any other name. Almost taking over the ship, the crew had filmed constantly inside the ship and on the flight and hangar decks. They brought six actors, the star being Strother Martin, who had played Liberty Valance's crazed, violence-loving gang member in *The Man Who Shot Liberty Valance*. Pete and I loved Strother ever since we saw the movie in '68 while waiting for a Navy charter from Norfolk to Oceana. To keep sailors from being bored and irritable and starting fights, the Navy crowded two or three hundred sailors, usually new recruits, into an aircraft hangar and showed a movie or two so they could pass the time while waiting for their specific Greyhound bus. In the military, soldiers and sailors were always waiting for something. I best remember Strother as the warden in *Cool Hand Luke* who addressed his chain-gang prisoner Paul Newman to complain that "what we've got here is a failure to communicate." I didn't see it until it came on television in '73, after my Navy days.

We cruised around in the eastern Mediterranean until we slowly began heading west. We anchored off Cannes on the Riviera on the 24th of June for three days of liberty. Pete and Gabe and I went into town to drink some beer. Pete met some American college kid there, and he wouldn't leave her, so we had to leave him. We saw him the next morning as he entered the AQ shop. He was very hung over and looked real bad.

"What have you been up to, Pete? You just getting in? Where'd you sleep?" I asked.

Pete could barely croak hello to Chief Cranston and hungrily chugged a cup of hot black coffee, ignoring me.

All our coffee was black because we didn't have a refrigerator for milk or cream, and we had no powdered creamer or sugar because it wasn't government-issue, and cleaning all that crap up would have been too much trouble anyway, if it had been. We didn't even have a coffee pot; we had to send a runner down to the chow hall two or three times every day to get a large pitcher full of piping-hot coffee and bring it up to our shop. We mounted a platform a little over head-high against the shop wall to hold a large container that had a faucet on its face, and we poured that chow hall coffee into the container. We didn't know if it was fresh or three days old; age didn't matter, it was coffee, and it was hot, boiling hot. When we got home from our cruise and took the container and the platform down to wash them, we found several large spiders that were as dead as Naples on Saturday night floating on top inside the half full container. We hadn't kept a lid on the container because we thought nothing could get in it; we'd never seen a spider in the shop before. Or, thank God, a big, filthy roach.

Pete followed me below to try to eat some breakfast before the chow hall closed. I hoped he wouldn't get sick. I asked him about the college girl he met the previous night. He told me he had snuck into the girl's room, and they ran her friend off. The girl pulled out some marijuana and some other stuff that they tried. They got sky high, and she wanted to make love.

"Man, I tell you. I was a little scared. She was so ready, and I was, too, but I began to worry about who she had been with and all that stuff. She was just too hot, and I didn't know if it was me, or just anybody, she wanted. She slipped off her shorts and top and jumped on the bed. I saw a little red rash on the inside of her thigh, but I said to hell with it and joined her. We took a couple of other hits after that and then we did the 'bad thing.' She was something else. As we were lying there, her roomie came back. We were both still naked but the girls didn't care. The new girl, who was pretty good looking, walked over and took a hit off my joint. I was staring at her cute figure and thinking how much fun she would be. She went over to the desk, pulled out a travelogue and started reading. We were there in the bed naked, and she was planning the next day's itinerary about six feet away; I couldn't believe it. You should have stayed with me; it was like being back in Olongapo."

"Pete, you'll never change. I've heard you say that before; 'you should have stayed with me.' We'll be home in a couple of weeks, and you know I wouldn't do that even if I wanted to; I'd be afraid I'd catch something, and with my luck, it would be something deadly I'd give to Jane."

"Oh, yeah, I forgot you are the married man now, a virtual saint."

"Give me a break, man. So, I'm married. Quit throwing it up to me. One day you'll be married, too. So where've you been all night? Screwing?" I asked.

"No, not all night. I was high as a kite, but I knew to get back to the pier before the liberty launch quit running. I got dressed, said goodbye and headed down the hallway. Everything was fine until I got to the front desk. The old man looked up from his magazine and asked me, 'Where'd you come from? Who are you?' I said I was just saying goodbye to my girlfriend from the States. The old man said it was after midnight, and I wasn't allowed to be in the hotel without an assigned room."

"You knew that," I said.

"Yeah, I knew that, and he knew that. I got by him going in, but he got me coming out. The old man grabbed the phone and said he was calling the police. He thought I was a robber or a rapist. I ran past the desk and out the door. I ran all the way down the boulevard along the docks. It was real late, but a few people were still coming out of the casinos and hotels and restaurants, and I kept running. They must have thought I was a robber, too."

"This was along the waterfront? The Riviera?"

"Yeah, I stopped a couple of times to catch my breath, but then kept going. I must have been three miles from our landing. When I got there, no one was around; I guess the launches had already quit for the night. I was afraid I'd be written up for being AWOL again. I sat on a bench and tried to think what to do, but I had to wait until morning. I found some boxes on the sidewalk and pushed three or four of them flat and sat in the middle. I fell asleep, I guess, and got up when I heard people talking. No one had come ashore yet, but a merchant was heading out to the ship with a load of groceries. The guys saw my uniform and offered me a ride, and I climbed aboard."

We left Cannes on the 27th in the morning and arrived at Spanish Naval Air Station Rota at Cadiz that evening. We had gone from the east coast of France to the west coast of Spain during the day. The ship clanged her bells and tooted some horns as we had passed through the Strait of Gibraltar. We anchored there for two more days while Change of Command Ceremonies were being held for the top brass. The U.S.S. *Saratoga*, the infamous *Sorry Sara*, was relieving us as the superpower's super warship in the Mediterranean Sea.

The *Forrestal* headed straight for home. She kept up a good pace at all times, day and night, except when she needed something. Halfway across the Atlantic, she had to refuel and take on oil to keep her boilers going. We slowed down and watched the oil tanker come alongside. The giant ship was about half as large as we were, but her four smokestacks and booms and superstructure loomed above the catwalks on our flight deck. She wasn't like a tanker you see off the beach while on vacation; she was a Navy warship that looked like a gigantic destroyer. The *Forrestal* and the tanker had a system which let the tanker's navigational gyroscope be coordinated with the carrier's. The two ships were then guided by the same gyro. That was a safe and sure system that kept the ships from colliding with one another. Nothing could go wrong.

As we connected with the tanker on her port side, one of our destroyers cruised up and connected with her on her starboard side. Soon the three mighty ships were cruising at the same speed side by side each about twenty yards apart. One of our crewmen picked up a large rifle and shot a line like a ski rope across the water to the tanker, and the destroyer did the same thing. Their guys picked up the weighted end of the rope, hooked it to a wench and began pulling it across the water. After crossing the twenty yards of sea, the ski rope was replaced with a heavier rope that had been tied to it. That heavy rope was then replaced with an even heavier rope tied to it, until at the end of the routine, a hawser the thickness of a fat man's thigh connected the tanker to the *Forrestal*.

There we were, just yards from the tanker and doing ten or so knots, connected by four hawsers each two-feet in diameter. The tanker was connected to the destroyer on her other side as well. It was a pretty day, and the seas were fairly calm but that big old tanker, as big as she was, was still slightly moving up and down in the sea. Pete and I were on the hangar deck watching the show from behind the elevator and getting ready to go up top. It looked like we were standing on a pier with a ship on the water moving up and down beside us. Between each hawser a five-foot-diameter oil pipe had been pulled from the tanker to us and tightly secured. Each of the three pipes was pumping oil under intense pressure into our ship.

Normally, when the refueling at sea was complete, the oil was stopped, and the pipes were closed off and returned to the tanker. The hawsers were then rolled back onto the carrier, followed by the next sized rope, followed by the next, until the little bitty ski ropes were pulled free. At that point the gyros of the ships would be separated, and the ships would be navigating on their own. All our escort destroyers had been refueled earlier that day except for the one refueling at the same time we were.

That was the way everything was supposed to happen, but this July 1, 1971, the next to last day of our deployment, on our way home with nothing to worry about, the safe and sure system failed. The tanker's crew disconnected her nav system and gyro from the *Forrestal*'s before the oil had stopped flowing and before the hawsers and pipes had been disconnected. It was a terrible, possibly deadly, mistake. Thank God, the escort destroyer had finished refueling a few minutes earlier and had peeled away. Pete and I had reached the catwalks beside the flight deck and slid underneath the tails of some Phantoms to get a real close look at the everyday operation. We didn't notice anything was wrong, but we did notice the front of the tanker was trying to drift away from us. The ropes and pipes were soon stretched tight between the ships. The Captain started barking commands from the bridge over the loudspeaker, and some bells and gongs were set off to advise our crew of a potentially bad situation, maybe a collision. The crews of both ships frantically began trying to uncouple the hawsers and pipes. The three connections toward the rear of the ships came free and dangled in the water. The emergency release on each of them had functioned properly and may have saved some lives. The front hawser and pipe were apparently under too much pressure to release. They continued to hold us together in that deadly encounter.

The tanker slowly curved away from us. The bow of the ship tried to move to the right, but the first set of ropes and pipes toward the front of the ship held firm. The Captain ordered our men to "Cut the line, cut the line." Surely we had torches on board that could have cut them or burned them, but I knew that no one would dream of burning anything around all that loose oil. The movement to the right at the bow of the tanker forced her rear end to move closer to us. The Captain saw what was going to happen and yelled over the intercom to us, to me and Pete, "Hey, you two men, get out of that catwalk. That's an order. Now."

We ran to the steps and climbed up and out of the catwalk and looked around to see where next to go. We stood on the flight deck, its starboard edge facing the tanker and crowded with planes, many with their tails hanging over the catwalks and the ocean. Five short, rapid blasts of the *Forrestal*'s shrill whistle warned everyone of an imminent collision between our two ships. Her internal automatic alarm system wailed "Collision. Collision." over its speakers, sounding like a programmed robot. We were standing there next to some F4Js when the smokestacks and booms of the tanker began to push against those planes. The metal skin of the tanker's body cracked and popped off like a naval orange peeled in one motion. The catwalk was crushed, and the tails of our planes were destroyed, crumpled like wadded-up writing paper. The noise of the destruction of the ship's side and the planes was horrific, and General Quarters was set, the gong, gong, gong of the PA system deafening. Visions of the 1967 *Forrestal* fire ran through my head. When I saw the tails of those Phantoms get crushed, I was afraid that the ship might catch on fire. Thankfully, there was little, if any, fuel left in those planes; they were empty because they wouldn't be flying again until we were closer to home. But after this accident, they didn't need any fuel for a long time.

Pete and I ran and hid behind a big, yellow, flight deck forklift used to move planes, and we waited. I wanted something between me and any explosion that was coming. The hawser and its pipe were finally cut through, releasing us from the tanker. The tanker lurched and rolled to starboard and then slowly continued on her way, her tail almost touching our tail. We secured from General Quarters when the small amount of fuel on the flight deck was completely washed overboard. Pete and I helped man a firehose washing jet fuel overboard. Everything was checked out, and the "All clear" was sounded. Someone caught a hell of a lot of grief for that crazy, costly mistake. Four men had already died during this peaceful goodwill cruise, and thank the good Lord, the number was not higher. Those brave souls, remembered in our cruise book, were AME2 James E. Scharf (RVAH-7), ABH3 Robert M. Moore (USS Forrestal), EM1 Wilbur B. Watson (USS Forrestal) and LCDR Richard Rutkiewicz (USS Forrestal).

On July 2nd, the *Forrestal* rounded Sewell Point in Hampton Roads, Virginia at around 1300 hours, or one o'clock in the afternoon for you landlubbers. The giant tugs quickly pushed her to Pier 12 at the Norfolk Naval Ship Yard, and my military career was over, just like that. The Navy gave me a break, they let me out a month early. Jane was waiting for me, and after we shared some heart-felt hugs and kisses, we said goodbye to Peter Boy and drove to our cozy little love nest at Oceana. We spent the night, and spent the night, and spent the night, then checked out of base housing the next morning and took off for Augusta to see my folks. Pete needed a ride, so he sat up front as navigator, and Jane sat in the back to be nice.

My W-2 earnings for my last half year in the Navy totaled $2,752. All military services offered a Variable Reenlistment Bonus in '71 to induce highly skilled specialists like Pete and me to reenlist. The maximum reenlistment bonus for sailors like us who served four years and were E5 AQBs was $3,125 a year for an immediate lump sum payment of $12,500 to serve four more years. We were both offered $11,000 in cash and four years of college tuition, and we both turned it down. It wasn't going back to Yankee Station in the Gulf of Tonkin that bothered me. On the contrary, I greatly enjoyed it, but I had gotten married during my enlistment and wanted to go home and start my family. I didn't want to be away from my wife half the time. Pete left because he discovered he wanted to live a wide open life unfettered with rules. He had begun to despise the Navy's and everybody else's rules and regulations. As I look back on it now, Pete would have been much happier had he stayed in the Western Pacific with all the booze, the bargirls, and the drugs he could stand. But in the final analysis, we may have left the Navy for the same reason Lt. Calley had left an insurance investigator job five years earlier in Albuquerque: we all felt we had reached our own personal levels of incompetency. Like Calley, Pete and I did the honorable thing: we quit.

21. THE DEAD AND THE QUICK

I will never forget the good times Pete and I shared those fifty years ago, the women, the booze, the excitement. It's not like I make an effort to think of him; it's just the mention of someone named Pete on the news or in the paper brings him back to me, like an old song from high school can sometimes make you think of someone special. I received a call asking me to come see him, and it worried me. I wondered why people kept associating him with me after all these years.

The stylishly-dressed young man at the funeral home walked me to the large prepping room in the back to identify Pete's body. I inched closer to the slab where he lay, flat and cold in death, with purple lips waiting for a goodbye kiss. The man touched me on the elbow and said the body was in pretty bad shape, and it wasn't really necessary that I view it up close.

He quietly, almost apologetically, said, "You don't really have to look at him, you know. Don't you want to remember him as he was?"

I said, "No, thanks. I'd like to look at him one last time if you don't mind. I'd like to see if he's really dead. I knew him for fifty years; I'll never forget the way he was, tall and good-looking. Smart and witty, too. At one time he meant something special to me. He could've been anything he wanted, but instead, he ended up as one sorry piece of crap."

The young man took his hand from my elbow and straightened up. "Oh, I didn't mean to upset you. I was just pointing out how bad his flesh is. He's been dead awhile. We're usually finished with these people a lot sooner than this guy; sorry, I meant your friend."

I wanted to be nice, so I tried to put the boy at ease. "Hey, it's fine. I don't get offended by anything anymore; I'm too damn old. Relax, it's cool. I've seen so much lately that I've become cynical; it takes a lot to stir me up. I believe you reap what you sow, and this dead piece of flesh looks just about right for Peter Boy."

"I'm sorry," the boy said.

"It's funny because at one time…years ago…he'd been a big wheel in the insurance business. Lots of money. Lots of drugs. But you know, the higher they fly, the harder they fall."

"Yeah, but the harder they fall, the higher they bounce," he countered.

"Touché. That's good; I haven't heard that before, but it makes sense."

How old and fat Pete had gotten. I was shocked when I saw his face; how much he had aged since we had parted ways. The young, good-looking ladies-man of his fast-track days was hard to recognize beneath the white, nasty flesh covering most of his body. I stood amazed, realizing the same thing had happened to me: staring at me in my mirror every morning was a bloated, splotchy face covered with knots and bumps and black cancerous spots. But Peter Boy's face was white and shriveled like it had been sitting in a bucket of water all day, and little chunks of rough skin were missing, revealing the pale pink flesh underneath. I flashed back to gawking at the burned sailor in the infirmary passageway and then his magnificent burial at sea. He and Pete had ended up the same, dead and alone and unrecognizable. But who in the hell was going to bury Pete?

Pete's appearance at this stage in my life didn't make any difference to me. He was Peter Boy, always had been Peter Boy, and would always remain Peter Boy until I died, too. His given name was Freeman Irby Gilbert, Jr., and I don't have a clue who started calling him Peter Boy, or why. But I liked it; I liked the sound of it. It was a much better name than Freeman, or Irby, as his father was called.

I had no feelings of sadness or remorse over his death. We're all going to die someday. Seeing him cold and lifeless on that slab brought back countless memories. I could name so many times he had publicly embarrassed himself and his family and friends after hard drugs had grabbed him by the throat and choked him nearly to death. But when I weighed the good he had done for his charities and his employees versus the bad things attributed to him, maybe he came out on top. Not by much, mind you, but just a little, just enough. But with the crazy existence he had created for himself, I was surprised he had lived this long.

When I had walked into the parlor, the funeral director shook my hand and said, "Thanks for coming down. We need someone to identify Mr. Gilbert's body for the record, a responsible person who knew him personally, and his attorney said you were probably the closest person to him. He said you would do just fine, but it took him a while to find you. He didn't want to see Mr. Gilbert again, but thought you might."

The director had turned me over to his son, his assistant, who led me to the back. The young man pulled down the tarp-like sheet covering the mound of flesh on the slab and asked me, "Is that Mr. Gilbert?"

"Yeah, that's Mr. Gilbert, or what's left of him. You know, my friend, I believe you reap what you sow in life, and this dead hunk of flesh looks just about right."

"We have some things we removed from him when we brought him in. Here…in this bag." He handed me a small zip-lock bag with "GILBERT" written on it with a black marker. "It's his watch, a couple of rings and a necklace. He must have been a ladies-man to have all that gold jewelry. But we didn't find a single piece of paper money on him. He either didn't have any money, or someone got to it before we did."

I opened the bag and slid the contents on the table next to his slab. I picked up his Seiko watch. The boy didn't know any better; he thought it was real gold, but it was just nicely made to look like gold. It was still running and had the correct time. I'm sure Pete bought it at a pawn shop when he used to have enough money to buy some jewelry, back when he cared more for material things than for drugs. The two rings were real and expensive. One was a three-dollar gold Indian Princess coin mounted in a very masculine setting. The other was a little pinkie ring he had been given by his first live-in girlfriend at Georgia. The gold necklace was heavy and thick and also expensive. It looked like the pimp he wished he could have been. I thought how honest these folks were to give this jewelry to his beneficiaries without stealing it. But it did look to me like he should have had a little money on him, enough to buy a burger or some cigarettes or some weed. But, maybe not; the assistant said no wallet or identification was found on Pete or in the motel room. The police finally located Mr. Cotter, his old corporate attorney, from some trust documents strewn around his room.

I got lost in a memory of long ago when Pete and I used to throw parties at our apartment on 63rd Street in Virginia Beach out by old Fort Story. In those days, we only drank alcohol and never touched drugs. We'd drink up every last drop of everything in the apartment and smoke every cigarette in the place before we crashed. In the morning we'd wake up hung-over, hungry, maybe a little sick, and immediately start digging through ashtrays until we found a butt or two to smoke. A cup of coffee the first thing out of bed was great, but nothing ever beat that first cigarette in the morning. Especially with that hot cup of coffee. Our hands would be shaking, and it would seem like it took forever for the coffee to brew. We'd have to hurry to the store for more cigarettes.

"Mr. Smith?"

"Oh, sorry," I said. "What's next? Tell me, I've never done this before. Whatta I do?"

"Nothing, really. The attorney told us to ask you if you want his ashes. We're going to cremate him."

"His ashes...will that mean I'll get stuck with the bill?" I was glad to take his ashes, but not if I had to pay for the cremation. Pete didn't mean that much to me anymore, and I had already spent too much on him bailing him out of trouble over the past five years. He could never repay me now; he should have saved some of his inheritance when his mother died.

"Oh, no, the attorney will pay for everything. But he wanted you to have 'em if you want them. He knew y'all were close. He even said for us to give you his personal things. He said you'd like them."

"You mean the jewelry in the bag's for me?"

"That's what Mr. Cotter said."

"Well, thank the Lord. A little payback after all."

I knew Pete wanted his ashes scattered on his mother's grave, so cremating was the perfect thing to do, but I didn't say anything to the boy. I was tired and ready to go, so I didn't start another conversation about that. I asked him what was next.

"Just sign these papers saying you'll take the ashes and have them properly interred." I signed the papers with the full knowledge I wasn't going to have them interred anywhere; Pete and I were a lot alike, birds of a feather and all that.

A week later, I got a call saying Peter Boy was ready to pick up. The funeral home had turned that three hundred and sixty pounds of morbid flesh that was once Peter Boy into a twenty-five-pound bag of burned bones and ashes. The container that held his ashes was a bright yellow plastic cylinder that looked like a hat box covered with tape from a crime scene. He was so heavy I had to pick him up with both hands. They crammed him in, so completely filling his container that they had to double tape it to keep it closed. The bright yellow bottom of the bulging container and the shiny black tape wound tightly around its neck irreverently reminded me of something I'd read.

"His cup runneth over," I mumbled, and the director heard it.

"Did you say something?"

"Oh, nothing," I said and thanked him and signed the release.

The director told me as I was leaving, "He was such a big man that some of the ashes on top are actually small pieces of bone that didn't burn up all the way. That's normal for a large person; we don't run them through the furnace twice. Just in case you look at them."

Jane had waited for me in the car at the funeral home. She didn't want to even think about Pete's cremation, or anybody else's, much less see anything concerning dead bodies or ashes. We took Peter Boy to our home for the last time. I remembered I once said he'd never spend a night in my house if he were on drugs, but I thought that maybe he was harmless this time. I called the pastor of our local Methodist church, and he agreed to meet Jane and me at the church chapel on Thursday to hold a little service for Pete. I arranged to meet my family at California Dreaming in Martinez for an early supper afterwards.

The Reverend prayed for Pete's soul, and the three of us offered prayers to remind the keeper of the Gates that at one time in his life Peter Boy had been a fine human being. Well, 'a fine human being' may have been overstating it a little, but the chances of his entering the Pearly Gates seemed to me to be at least a fifty-fifty toss-up. Maybe not. The Reverend had given Peter Boy a nice little service, and I handed him an envelope containing a brand-new hundred dollar bill as an honorarium for his troubles. He had done a good job, and I paid him what I hoped was sufficient for a six minute service. I put Pete in the backseat, but I didn't connect his seatbelt. I was going to take it nice and easy, and I didn't think he was going anywhere.

My family last saw Pete right after we had gotten out of the Navy, back when he was worth knowing, and I wanted them to say goodbye to him along with Jane and me. None of them was close to him, but I hated to think of someone being buried with only one or two people present. I had seen many pictures of homeless men and women and unclaimed bodies buried that way, and that always seemed so sad and lonesome to me. As I sat around the table at the restaurant with my wife, my mother, and one sister and her husband, I remembered the good times we'd had with Pete and his wife in Swainsboro and Augusta and Atlanta.

We took Pete's ashes to his mother's grave at Westover Cemetery and sprinkled them over her resting place. Ten years earlier, he had asked me to do that for him when he died. The idea had come to him while he was in prison for mail fraud and embezzlement. He had heard through the grapevine that prison land had gotten scarce, and in the future, prisoners were going to be buried standing up. The thought of him standing on his feet the rest of eternity had him crying like a baby on the phone. Before our little service, I had asked my Aunt Ann, who worked at the cemetery, if we were allowed to sprinkle his ashes there. She said, not really, but we all go home at five o'clock, and the gates don't close until dark. We held hands in a circle as we gathered around the grave with its very thick coating of fresh ashes. Some of the larger pieces of bone had rolled off the pile, so I stuck my foot out and pushed three or four of them back on top. I hoped it would rain before the ashes killed the grass. Jane prayed for Pete's salvation and sobbed softly as she added to her prayer, "maybe we didn't love you as much as we should have." But we had at one time, more than we should have, before it became too much of an aggravation.

Three weeks passed before Mr. Cotter called me with the news that Peter Boy's estate was settled. "Come on down to my office and get a key, and you can have whatever you want out of his motel room. I don't think there's much. You know we sold his mother's house and gave him the money. It lasted nine months, about what I thought; then he lived on Social Security and anything else he could scrounge. His last couple of years were a far cry from his glory days in Atlanta."

At his office on Greene Street, Mr. Cotter handed me the key. He told me what room Pete had been living in on a long-term lease and told me to take any of his belongings I wanted. I drove to the Scotland Arms and parked close to the building, which was just one long line of rooms, three stories high, fairly new and freshly painted pearly white. I cut off the engine and slid out of the car. It was deathly quiet. The clanking from the closing of the car door enticed two women to stick their heads out of a room a couple of floors above me. Maybe they thought I had come to put one of them, or both of them, to work. It was two o'clock in the afternoon, but in their business, it was never too early or too late. They were like a mega-store; they never closed and they were always ready.

"Whatcha want, big boy?" one yelled down to me.

I shouted up at her, "Where's room 335?"

"It's right up here, sugar. Down that way a couple of rooms. Whatcha want?"

I said, "My friend's dead, and I've come to clean out his room."

"Oh, no, your friend's dead? The big fat guy? Santa Claus? I'm so sorry; he was so good to us."

Another gal edged to the upstairs railing and leaned against it with her legs pressed tightly together. I looked up at the woman at the railing and under her short skirt, it seemed she was not dressed. I assumed she was aware of her predicament and was just doing a little bit of advertising. I guessed she knew I wasn't a cop because I was a friend of Pete's. "Come on up. I'll see if the door's locked."

"I've got the key," I said.

I walked up the two flights of stairs and was surrounded by the two ladies of the night, except these were also ladies of the day. They swarmed over me and told me how much they loved Pete. He would give them the shirt off his back, they said. And if he had a dollar and you needed it real bad, he would give it to you without a second thought. He was always bringing them hotdogs and hamburgers. They giggled when one of them said, "Yeah, and sometimes he brought us some good stuff."

I'll bet he did. But I didn't say that to them. They were very pleasant and helpful. Peter Boy's room was just another cheap motel room with an unmade bed. He had slept on it so many days that the middle sagged a few inches due to his great weight. The bedspread was dirty and had yellow stains dotted all over. The pillow was missing its case and was striped and dirty like one found in old southern prison camps. He had ten or twelve jars of chow-chow and other condiments neatly lined up on the dresser but only two were unopened. I took both of them. He was a big, big man so I looked through the giant shirts and pairs of pants in the closet just for curiosity. His huge white overcoat was so large he could have been the Grand Wizard at a cross burning without putting on an official robe; I think now that maybe he was. I held the coat up to eye-level just for fun, and it still touched the floor. Somebody else could have it all.

"Who was here when he died? Do any of y'all know how he died?"

"I was here. I was right next door."

"Well? Tell me what happened; I've been his friend since high school."

The woman sat on the corner of the bed. I would have sat on it, too, but I was afraid I'd catch something. I hadn't changed much; I was still a lot like I was in my Navy days. The room smelled sour like spilled Clorox or rancid trash or sex. There was even a pair of soiled panties on the floor under the edge of the bed. The shameless way the woman was sitting made me think they were hers. I flashed back to a sixty year old Filipino mama-san at the West End Club in Olongapo squatting beside me at my table, a betel nut bulging out her cheek, and flashing me her one black tooth.

"Priscilla was here. She told me what happened. I wish she was here to tell you herself, but they took her down to the police station, and she ain't got back yet. She and your friend had been watching TV when there was a knock on his door. She was supposed to be off that night, but the man at the door was a regular, and she liked him, so she went to her room with him."

She fidgeted with her cheap bracelet and lit a cigarette. "Want one?" she asked.

I shook my head no, and she said, "Your friend, we called him Santa Claus 'cause he's fat and used to have a big white beard, he was fine for a few minutes when she first left, and then he got mad and pulled out his gun and was flashing it around. He was in a rage and kept yelling, 'You don't know who I am. You can't do this to me. I have friends who'll blow your brains out. You don't know who you're dealing with.'" She got more excited the more she told the story.

I liked that: Pete's pulling the gun out and flashing it around. I was a gun collector, and he had one he wouldn't be needing any more. Maybe I could get it; it should be mine anyway, according to Mr. Cotter. "Where's the gun?"

"The police took it with them."

"I'll be damn; the police have it. Go on…please," I said. I wasn't going to go to the trouble to try to get it from the police; besides, it was probably stolen.

"I heard him bumping against the wall outside my door and muttering. He had been doing meth during the day and was feeling kinda rough. I got him to come back inside and smoke a joint with me until Priscilla came back. That worked okay. After an hour or so she came back from doing that guy, and she and your friend started on some more drugs. That settled him down because she went back to his room and banged on the wall and yelled to us, 'Come on over, I've got the money; let's party!'

"Your friend drove down to the Sans Souci and bought us a party bag with Priscilla's money. He came back, and we did a lot of stuff 'til two-thirty. We were all tired and dirty, so we split up and went to bed. Priscilla stayed with him. Anyway, I heard a scream the next morning and ran out of my room. There was a policeman at your friend's door. Priscilla was barely awake and was sobbing and carrying on like something crazy. I looked inside, and he was lying there naked on the bed, all fat and white. He had a bluish tint in some places and on his fingers and toes."

"Jeez, that's scary. Ever happen to y'all before?" I asked.

"No. We've seen some crazy, mean things, but this was a first. Dead and cold and blue."

"Sorry; please go on."

"So I asked what happened, and the policeman stretched his arm across the open door. He said, 'Stand back, lady, this is a crime-scene.' He looked at me again and could tell I belonged there and put his arm down. He said he had come to check on Priscilla because she had missed a meeting with her parole officer. When he knocked on her door, nothing happened, so he asked one of the girls outside where she was. Somebody told him to try your friend's room. The cop knocked on the door and tried the handle when no one answered.

"The door was unlocked, and he pushed it open. He saw Priscilla and your friend on the bed. He thought they were sleeping and shook Priscilla awake. As she tried to figure out what was going on, the policeman tugged on your friend. He tried it again and stopped. Your friend was cold as ice, he said. He was dead and had been all night. That's when Priscilla rubbed her eyes and saw that he was dead and screamed."

"That's all we know," the other prostitute added as she stuck her head inside the room. "He was a good man to us. We don't know what got him down so low, but we thought he was great. He told us some good stories about when he was growing up and when he was going to college and in the Navy. You his Navy buddy? Smitty? He talked about you all the time; he loved you."

"Yeah, well, he had his good days…Like all of us. We'll all be cold and blue someday, just like he was…He was my best friend for a long time; we were real good buddies. We saw the world together. Thanks, ladies," I said, as I grabbed the two jars of chow-chow, "I'll be back to see you some time."

I laughed, and they giggled.

"You come back, you big hunk. I'll let you have anything you want…for free," my scantily clad friend yelled down the stairs as I went to my car.

22. WHAT HAPPENED TO CALLEY

I continued following the life of William L. "Rusty" Calley in my spare time after I left the Navy. My sister-in-law would write to me and talk with me on the phone, telling me the stories she got from her momma and poppa. She sent me a copy of *Lieutenant Calley: His own story as told to John Sack*, duly autographed to her parents by Calley. The inscription reads: "To Jay and Louise, with many thanks for the warm and loving friendship, and the resurance during my entire ordeal, William L. Calley." The fact that Rusty misspelled reassurance did not surprise me; I think that was about par for him. It was just one small example of the faults and inadequacies which may have made him feel insecure and unworthy during OCS, while hiking through the jungle surrounded by many better-educated college dropouts who had chosen not to flee to Canada. I sat down a few months ago and finished, as far as I'm concerned, my Calley story:

From his early teens, anyone could see that Rusty Calley would struggle in positions of leadership of almost any capacity. His lack of good judgment, a character flaw easily excused in his youth, doomed him in his twenties when decisions of life or death had to be made. The circumstances of his childhood were like most middle-class white Americans in the '50s and '60s; his father, a World War II Navy veteran, owned a heavy equipment company and his mother took good care of Rusty and his three sisters. The family was not poor by any means; they lived in Miami but owned a cabin in Waynesville, North Carolina. They were totally normal, but Rusty often worried his father drank too much. A fairly content young man but possessing a quirky attitude, Rusty didn't take most things seriously, including his schooling. His problem wasn't in his environment, it was in his head; it was his lack of good judgment.

Caught cheating in the seventh grade, just like so many other teenaged boys, Rusty was made to repeat the entire school year. That harsh and unusual punishment for an almost common infraction would embitter and depress any child; it could make someone even want to kill somebody. After being expelled for a week in the ninth grade for arguing too long and hard with a teacher, a man who had called him a wiseass, he dropped out of public high school and enrolled at the Florida Military Academy in Fort Lauderdale. That move required paying for room, board and tuition and was obviously a decision his father made. A year later, Rusty transferred to the Georgia Military Academy in Milledgeville, but some force, most probably his father's bank account and dwindling patience, brought Calley back to Miami. He lived at home, tried to be sociable and graduated from Miami Edison Senior High School in 1962, ranking number 666 out of 731 students.

That ominous 666 raised the eyebrows of many of his Miami acquaintances because it is the sign of the Devil. That ranking also meant that Rusty was only 12 graduates away from being in the bottom ten percent of his graduating class. That in itself means nothing; many good people are in the bottom ten percent of their class. It is, in fact, a statistical certainty that there is always a bottom ten percent of every class and every other thing. But this unimpressive 19-year-old would soon be leading American troops on the field of battle in the midst of a war, leading men in battle, making life and death decisions, using his poor judgment. Jesus Christ, that can't be true.

Palm Beach Junior College in Lake Worth, Florida, accepted Rusty in the fall of '62, but he just wasted his time and squandered his daddy's money fooling around and failing most of his courses. Dropping out in '64, Rusty spent his time working odd jobs in restaurants, hotels and car washes. Unhappy with his existence, he tried to join the Army, but was rejected due to a medical issue, a bleeding ulcer. He didn't realize that the Army, like the religious hierarchy of the Old Testament, required unblemished, perfect specimens for sacrifices; no damaged animals were worthy to be killed, either in patriotism or the name of Jehovah. It was, however, worthy of him to consider enlisting; Vietnam had not yet escalated into the killing machine it would become, but it was already a very dangerous prospect in all respects. Maybe he remembered the inspiring words of President John Kennedy who told our young men at his inauguration to ask not what their country could do for them, but to ask what they could do for their country. Maybe Kennedy's assassination still meant something, even to a simple man like Calley.

In March of '64, just three months before Calley's enlistment attempt, Colonel Floyd James Thompson was captured in Vietnam after the plane he was in was shot down, making him the war's first American prisoner of war. Maybe Rusty saw Col. Thompson on somebody's television and vowed to join the Army to avenge his capture and free him, just like a Rambo 40 or so years later would do. Or maybe Calley just needed a place to sleep and food to eat that day. Rusty's salvation arrived in the midst of a union strike, when the Florida East Coast Railway hired him as a conductor. Unfortunately, he was arrested a few months later for letting his train block rush hour traffic in downtown Fort Lauderdale. He was soon cleared when it was proved the train's brakes had malfunctioned. Calley loved that job, but the strike ended, returning the union men to their positions, and he was once again on the prowl for a job.

During the course of world events, seemingly minor turns of fate often deeply influence other, later events. If Fidel Castro had been skilled enough to join an American baseball club as he often dreamt, there might possibly have been no Bay of Pigs or Cuban Missile Crisis. Had Adolph Hitler been able to make a living as the professional artist he worked hard to be, there might possibly have been no World War II or an attempt to exterminate the European Jews. Likewise, if Rusty Calley had been able to continue working at that railroad job he loved so much, there might possibly have been no My Lai massacre in Vietnam. But it was not to be.

As Rusty hopped around trying to make some sort of living, he received word his mother had cancer. She fought it as best she could until it finally won. His father was soon diagnosed with diabetes and couldn't work, and his business failed without him. The family home had to be sold, and the Calleys moved to their North Carolina retreat. To his credit, Rusty moved in and helped out as best he could, but true to his track record of personal failures, he became depressed and quit, leaving the care and support of his father to his three sisters. Calley bought an old Buick and headed westward. Stopping in New Orleans, he tried his hand as an insurance investigator. He realized while on an assignment in New Mexico that he didn't have the brains to do his job, so he did what was to him the honorable thing: he quit. That is actually the honorable thing for anyone to do when confronted with the truth that he cannot competently perform his job. Rusty wandered around out west for several months doing the odd jobs that paid his few bills. He didn't do drugs or drink heavily, but he smoked a little, so he had a small nut to crack each month. He was sociable and a little shy at times, but he wasn't a loner; he had a couple of ordinary, decent friends and an occasional girlfriend. He wasn't a ladies man, but he scored every now and then. He decided to move on to California where there were more fish in the sea, fish of many varied species, shapes and colors. Imagine being single and 23 and living in California in 1966. Nothing could mess that up.

While working in San Francisco that summer, Rusty received a tattered, many-forwarded official notice informing him the Miami Draft Board wanted to see him. A draft board 3,114 miles away wanted him to report to them for what? Induction? How many marginalized, draft-aged boys living on the edge would have responded to that letter? Many boys would have told their friends, "No way, man, I'm enjoying the good life, free as a bird, giving peace a chance. If the draft board ever finds me, I'll swear I never got the letter, so I don't even need to hide. If I act ignorant, what will they do? Send me to jail? No, man, they need me in Vietnam." But Rusty was going back to Miami. Was he that stupid, or did that decision measure the spirit, the essence, of the man? Did he believe in the American system that much that he was going to obey a letter that had taken months to find him while he was lingering in the midst of a world of hippies and drugs?

On his return to Miami to report as ordered, Rusty was stranded in Albuquerque with no funds and one sorry piece of junk that had died in the parking lot of a mall on the edge of town. As fate would have it, an Army recruiting station was sitting there, in the mall, waiting for his arrival. Looking for money to continue his trip back to Miami, Rusty thought maybe the Army would give him some. Well, that seems foolish, but he was genuinely heading to his draft board, and draft boards and the Army were connected, weren't they? The recruiting sergeant told him the Army didn't give or lend money, but why didn't he just join up and make the Army a career. Rusty said they had rejected him before and, besides, he didn't want to be sent to Vietnam. Staring at the handsome, squared-away clerk in the poster on the wall, he rubbed his chin and said he might be interested in a job like that. The recruiter knew he was hooked and reeled him in. He said he'd put him in for a clerical school, just like the poster boy, and guarantee him stateside duty if he'd sign right then. What did Rusty have to lose, other than his life; he had no money, no job, no place to stay, no way home, and the Army was looking for him anyway. The recruiter told him he was behind on his monthly quota and would loan him some money if he would sign now. Rusty knew he had no real future and decided to give it a try; he joined that July of '66, and Vietnam, his draft board and his old medical deficiencies be damned. Unfortunately, the Army had changed and overlooked his prior rejection; they needed new blood.

Thus begins the controversy of fighting a war requiring more good men than a country can easily provide from its citizenry. Recruiters were having to fill their quotas however they could. Our country was in the middle of fighting a brutal, unconventional war, an unpopular war. President Lyndon Johnson had increased the number of fighting troops in Vietnam to 292,000 in August of '66, and not enough red-blooded young men wanted to saddle up for a war few Americans whole-heartedly supported. Recruiters had to sign up whomever or whatever they could find. The sad result was that many men, armed to the teeth, were put in the field, untrained and unprepared for what they were to soon endure. Some men shouldn't have been there; Rusty Calley was probably one of them.

Getting started in this, his first real career, Rusty did his mandatory eight weeks basic combat training at Ft. Bliss, Texas, with no problems as a typical, enlisted boot camp puke. Nothing happened to purge him from the ranks. Near the end of '66, Rusty completed another eight weeks at the advanced clerical school at Ft. Lewis, Washington, as his recruiter had promised. But an amazing thing had happened; PFC Calley now found that he enjoyed what he was doing. He paid attention to his responsibilities, he squared himself away and took a deep interest in improving himself. The Army took notice and offered him the opportunity to advance. On his Armed Forces Qualification tests, Calley scored high enough to be considered for Officer Candidate

School (OCS), and he applied and was accepted. At OCS at Ft. Benning, Georgia, Rusty's work proved to be substandard, primarily due to his lack of command presence; he was only five foot three and baby-faced, making it hard for him, or any other man in that condition, to excel in looking and acting confident and authoritative. Besides, being substandard didn't always mean being unacceptable, especially for an Army at war. The moment William L. Calley, Jr. was transformed from a PFC doing superior work as an Army clerk into a future Army officer enrolled in OCS doing substandard work, he became the epitome of the Peter Principle by having been promoted to his level of incompetence. Rusty later complained that he was taught almost everything he needed to know at OCS, but nobody ever told him he would confront innocent civilians during a mission in Vietnam. "It was drummed into us," he said, "Be sharp! On guard! As soon as you think these people won't kill you, ZAP! In combat, you haven't friends! You have enemies!" Over and over at OCS he heard that, and he decided he would always act as if he were never safe in Vietnam, as if his life were constantly in danger from everyone and everything. As if everyone in Vietnam would do him in. As if everyone over there was bad.

In May of '69, in a jeep near the Chu Lai Base, Calley and two other Americal Division officers wearing tee-shirts passed another jeep carrying five U.S. Marine enlisted men who were horsing around and driving erratically. According to U.S. Marine military trial records, Calley's jeep pulled the Marines over and one of the Army officers informed the men that they had better square away. One of the Marines didn't like being lectured by a soldier and said, "We ain't soldiers, motherfucker, we're Marines!" He obviously didn't think the man had any authority over him, or maybe it was just a challenge. In any event, the three Army Lieutenants jumped out of their jeep to discuss the Marines' deplorable conduct. The Marines also jumped out of their jeep, and the Lieutenants quickly had their butts whipped. Calley was merely beaten up, but the other two officers ended up in a hospital for a short time. Either the Marines had gone easy on them, or the officers were pretty damn tough, because three against five wasn't a fair fight. Calley may have suffered a harsher fate had he been a bigger target or a better adversary. The Marines pleaded guilty at each of their special courts-martial and testified they had not known the soldiers were officers; to them it was merely an obvious case of mistaken identity which could have happened to anyone.

On June 5th, Lt. Calley was recalled to the U.S. from Vietnam, but he didn't know why. He foolishly thought they may be giving him some kind of medal. He was stunned when he heard that as a participant in the My Lai massacre he was to be tried in a military court of law. Calley said a colonel in the Inspector General's Office sent him to Ft. Benning, Georgia, as a favor, because it was the best fit for him; it was infantry through and through. That's where the trial would then be held. He was told all the jurors would be Army combat veterans, and all but two were. Calley would be judged by men who knew what combat was really like. At another fort, the colonel said, he'd be judged by WACs and doctors; people who may try to humanize the Viet Cong and dehumanize him, people who may make him out to be a greater monster than he was. The colonel, like the majority of the American military and the American public, was on Calley's side. Few people believed the troopers of Task Force Barker could have committed those crimes unless someone higher up had ordered it. Yes, they said, crimes had been committed but Lt. William L. Calley was just the scapegoat. He was the lowest ranking officer of the whole affair.

Warrant Officer Thompson was asked on June 13, 1969, to identify any officer he had argued with when he landed his chopper at My Lai, and he readily picked Rusty Calley from a lineup. One wonders where Lt. Brooks and Capt. Medina were; were they not in the lineup or had Thompson not recognized them. Or were they not identified by Thompson because he obeyed his instructions to the letter, and he had not actually argued with those two. In July, after being granted immunity, Pvt. Meadlo became the first GI to admit his guilt, and his confession was presented to the Inspector General. On August the 4th, when it became apparent crimes were committed, the investigation was turned over to the Criminal Investigation Division. Authorities at Ft. Benning decided to press charges against Calley on the 19th.

The personal photos of Sgt. Haeberle, the Army photographer at My Lai, were given to a court detective on the 25th of August, and with that damning first hard evidence, the court charged Lt. Calley on September 5, 1969, with six counts of premeditated murder in the deaths of 109 civilians. A conviction could mean the death penalty for Calley. Five days later, NBC first broke the news that Calley was accused of killing some South Vietnamese civilians a year earlier.

Two months later, in November, reporter Seymour Hersh's interview with Calley at Ft. Benning was published in many newspapers. The papers' follow-up investigations quoted the survivors of the massacre as saying over 567 civilians were killed. The General of the Army, William C. Westmoreland, ordered the Army's official investigation of My Lai to begin. Over the next week, the inquiry named 10 suspects responsible for the killings and started taking depositions. The world was stunned when Sgt. Haeberle's photos were published. President Nixon on the 8th of December in '69 said that a massacre apparently did take place, but it was only an isolated incident. Nixon was either told that by our military leaders or said it on his own to protect the reputation of the American GI or, more probably, the reputations of the past three political administrations. A week or so later, 45 members of Charlie Company were found responsible for crimes which ranged from violation of the rules of war to murder.

Interviews and testimony were taken through March 7, 1970, and Captain Medina was charged with assault with a deadly weapon and the premeditated murder of over 100 civilians. Another 25 men were charged with crimes on the 15th. Lt. Calley's court-martial began on November 17, 1970, eight long months later. The prosecutor said Calley had ordered his men to deliberately murder South Vietnamese civilians. Calley said that he had just followed Captain Medina's orders.

Another GI on trial for murder was Sgt. Charles E. Hutto who had admitted to investigators he fired his M-16 into a crowd of My Lai civilians. His defense at Ft. McPherson, Georgia, was simply that he was following orders. The Government provided no witnesses who could say Hutto actually hit any of those civilians at whom he shot. The defense team told the jurors, all combat veterans, that if Hutto had killed any of those civilians, it was not his fault. The fault lay with an Army that had failed him by placing him under the command of an officer who gave him an unlawful order. The GIs of Charlie Company had a moral choice and a legal responsibility to question unlawful orders according to Army Field Manual 27-10's Law of Land Warfare, but its officers were held to an even greater standard of duty in understanding and enforcing the laws of war. The U.S. Manual of Courts-Martial says a soldier must disobey an order that a man of ordinary sense would know is illegal. Any serviceman doing so must have balls the size of oranges, because he is immediately considered guilty of violating the law of "instant obedience" and has to prove the order was illegal at his subsequent court-martial. The March 8, 1971, Newsweek magazine quotes Calley as testifying that the Army taught him that all orders were presumed to be legal and all had to be obeyed. He testified that he was taught that "everyone was a potential enemy" and "because of the unsuspectedness of children, they were even more dangerous." Hutto's not-guilty verdict crushed the prosecution because the Government feared it would set a precedent favoring acquittal if a GI was diligently following orders, and there were 10 more My Lai cases to try.

One of Calley's most damaging witnesses was Pvt. Dennis Conti who testified that Calley told PFC Meadlo and him to kill the civilians, not guard them, on the road south of the village. Under cross-examination designed to diminish his integrity, Conti was asked if it were true that he had been smoking marijuana the day of the killings. He was asked if he had wanted to punish Calley for pulling him off the woman he was raping in the hamlet. He was asked if he held a pistol to a 4-year-old's head as he made its mother perform "an unnatural sex act" on him. The defense asked everyone in the court if anyone could believe that man. Testifying in Calley's defense were 21 members of Charlie Company who corroborated Medina's orders to kill everyone because they were all Viet Cong. Calley said, "I was ordered to go in there and destroy the enemy. That was my job that day. That was the mission I was given. I did not sit down and think in terms of men, women and children. They were all classified the same, and that's the classification that we dealt with over there, just as the enemy. I felt then and I still do that I acted as I was directed, and I carried out the order that I was given and I do not feel wrong in doing so."

The six-officer jury, after deliberating for 79 hours, convicted Lt. William Laws Calley, Jr. on March 29, 1971, of the premeditated murders of 22 South Vietnamese civilians. He was sentenced on March 31st to serve a life imprisonment at hard labor at Fort Leavenworth, KS in the U.S. Disciplinary Barracks, our only military maximum security prison. Of the 26 officers and soldiers charged for their crimes at My Lai or its cover-up, Lt. Rusty Calley was the only one convicted. After the conviction, the White House received over 5,000 telegrams, 100 to one in favor of leniency for Calley. In one poll of the American public, 79 percent disagreed with the verdict, 81 percent believed the sentence was too harsh, and 69 percent said Calley was just a scapegoat. The public was outraged not because Calley was convicted, but because he was the only one convicted. According to a Senate report, over 325,000 South Vietnamese civilians, all of whom were men, women and children, had been killed from 1965 to 1971, the time of Calley's trial. Most of those deaths were due to allied action and most resulted from massive aerial and naval bombardment and artillery barrages. Calley's defense said the American public wanted to know who, if anyone, was convicted of those war crimes. President Nixon had Calley removed from prison three days later on April 1st, just coincidently April Fools Day, and put under house arrest at Ft. Benning. The President's actions strengthened the case for the exoneration of the other officers of the My Lai incident. On August 20th, the commander of the Third Army reduced Calley's sentence to 20 years. Also that August, Captain Ernest Medina was acquitted on all charges. Calley's house arrest was served at his small apartment at 31 D Arrowhead Road in Ft. Benning. All types of alcohol, beer and liquor, were removed from the apartment, and it was manned by an MP at all times. He was allowed to leave the apartment, always under guard, only to eat at the mess hall or to exercise for an hour a day. His visits and telephone calls were limited to those on a court-approved correspondence and visitation list.

On February 27, 1974, after Calley appealed his conviction to the U.S. District Court for the Middle District of Georgia, Judge Robert Elliott granted a writ of habeas corpus and set Calley free on bail. As the Army appealed that decision, the Secretary of the Army Howard H. "Bo" Callaway reviewed Calley's conviction and sentence, as was required by law. Callaway reduced Calley's sentence again, this time to just 10 years. Under military regulations, prisoners are eligible for parole after serving one-third of their sentence, making Calley eligible for parole after serving three years and four months. The U.S. Court of Appeals reversed the district court's ruling, and Calley was locked in the stockade at Ft. Benning on June 13, 1974. He was released from custody September 25, 1974, when the district court ruled once again that he had not received a fair trial. Another series of appeals and reversals began which lasted for two years. During that time, Calley even appealed his conviction to the U.S. Supreme Court which declined on April 5, 1976, to hear it. Calley's conviction and 10 year sentence was reaffirmed by the U.S. Fifth Circuit Court of Appeals on September 10, 1976. On that date, Calley had only 10 more days to serve before he was eligible for parole. Secretary of the Army Callaway had previously made it clear that he would parole Calley as soon as his day of eligibility came. Therefore, the Army decided not to return Calley to prison for his last 10 days. Rusty Calley had served a total of only four months behind bars.

What was it like to be free on bail and not in prison, awaiting a trial in which you could be sentenced to death, during the other 62 months from April of '71 until September of '76? In his book, *Lieutenant Calley: His own story, as told to John Sack*, Rusty said he had read all the books and editorials written about his life and conduct and the incident at My Lai. He had read and studied thousands of letters and articles, most of those from the heartland of America, supporting him. He said he often got away from the ordeal of the constant scrutiny of the Government's investigations by going to cities like Atlanta and New York City. In the big cities, he went to fine restaurants, plays and sports events where he was never recognized or bothered. On one occasion, a girl in Greenwich Village stared at him until he was uncomfortable. He thought she was going to confront him about the war, but he said she turned out to just be a hooker, drumming up a little business. He didn't say if he had enough money to give "piece" a chance, as John Lennon almost wrote in July of '69. Rusty loved the big cities and blending in anonymously with the crowd (as anyone in his situation would).

The thing Rusty loved most as he awaited his trial was speeding down the Chattahoochee River in a ski boat and flying by Jay Skinner's ski ramp to send his fellow Army officers high in the air. The boys shared the fun and excitement with Jay's daughter Sue and her husband, my brother Jimmy, and their two kids. Sue was an expert water skier who developed her skill on the Chattahoochee at her parent's cabin on the Alabama side above Phenix City, across from Columbus, Georgia, and Ft. Benning. Jay's secretary had skied professionally at Cypress Gardens near Winter Haven, Florida, and encouraged Sue to sign a long-term contract with them. Louise told Sue she was going to the University of Georgia, so forget it. The ski ramp on the river (instead of Jay's cabin they always said "going to the river") was her specialty, as was Rusty's. Different groups of officers from Ft. Benning would come up on their days off, typically the weekends, and bring Sue's momma lots of beer to show their appreciation and fill the refrigerator. Louise loved to drink beer, and Jay enjoyed a glass or two of bourbon. Miss Louise also fed "her boys" whenever they were around.

Being in charge of the military business for a southeastern contractor constructing and maintaining military housing on five bases, Jay Skinner met daily with Army officers at Ft. Benning and other installations and had a strong affinity for military men. Some men he met and particularly liked were told how to find the cabin and its ski ramp and were given an open invitation to use them both. On one occasion, a couple of the boys came roaring in on a sleek ski boat with Rusty on board, possibly the one he owned. From that day on, Jay and Louise took him under their wing. He frequented the cabin from the time President Nixon released him from prison on April 1, 1971, until he was ordered back to the Ft. Benning stockade on June 13, 1974. The Skinners felt Rusty had suffered enough as he was scrutinized and investigated for following his orders and trying to do his duty, and they wanted him to feel like he had a family where he was accepted as an imperfect person. They felt the same for all military personnel, for good reason.

Louise was married to a soldier, Sue's daddy, at the beginning of World War II. He never drank while on duty, but once he was off, he usually got commode-hugging drunk. When she was six months pregnant with Sue, he began beating her when he was drunk. Louise finally had enough when unborn Sue almost died. A couple of Louise's brothers confronted him when they heard of that terrible beating and threatened him, sending him away. It was four months after Sue was born that the divorce was final. Louise had waited until after the birth so the baby would have her daddy's name. After a while, Louise ran into an old school sweetheart, and they began dating. He too was in the Army and was killed on Anzio Beach, Italy, in 1944. He left behind a baby daughter named Dianne, Sue's only sibling.

Louise met Jay Skinner after the war, and they married. She was the mother of two young girls and had been married twice before. Jay was a U.S. Navy veteran of World War II and had five brothers who all served our country during World War II or Korea. Jay was sent to Miami Beach during the war to train dogs for tracking enemy soldiers and fugitives, and as scouts and sentries, not just for Miami Beach but for all over. Some dogs were trained to locate booby traps, weapon caches, enemy fighters and snipers. Their keen senses of smell and hearing far exceeded anything humans could muster at that time before advanced electronics. Some dogs were used to defend camps or beaches primarily at night. Jay taught them to bark or growl at strangers to alert our guards on patrol. The U.S. Army had a dedicated dog training school in Ft. Benning and transferred Jay there.

Before the trial and when Rusty was not under house arrest, he was a regular at the river. He brought his live-in girlfriend with him often, but Louise wouldn't let them sleep together. They broke up after the trial; a stranger had come up one day as they were shopping and began to trash him and berate her for being with him. She turned her back on Rusty and walked away. He took her lack of support and disgusted attitude for what it was: her lack of support and disgust for him.

Louise and Rusty became quite close; she was his "Georgia momma." The Skinners were given passes to all the days of the trial, and

Louise made sure Rusty had everything he needed while he was under house arrest at Benning. In 1966, a correspondent named John Sack published his book *M*, the true story of a soldier of the U.S. Army's Special Forces unit called MIKE Force that fought in Vietnam. The journalist had primarily focused on the history of a long list of American war crimes and knew his subject. When the My Lai massacre was made public, he was approached by Harold Hayes, the editor of Esquire magazine, about writing Calley's story for Esquire. Initially declining to accept the assignment, Sack added, but "if it's about a normal person who's capable of doing terrible things, and it illuminates something profound about the human condition, then I'll do it."

Sack and Rusty talked for almost four months, as he recorded Rusty's own story in Rusty's own words. He stayed at the Skinner's river cabin whenever he was in town. He published his conversations with Rusty as a book and released it in a three-part series in Esquire. Calley said that after his story came out, everywhere he went people came up to him and told him how much they supported him. He received mail from all over the world, from one extreme, letters crying over the awful injustices done to him, to the other, letters wishing he would die and go to Hell. Four young ladies worked at Rusty's tiny apartment to handle the tremendous volume of mail and contributions he received. He was contacted by governors and the President, he was cheered and asked for autographs, and he was threatened with death. John Sack was indicted by the Government for refusing to turn over his reporting materials to assist them in prosecuting Calley. He said, "I am a journalist, not an agent of the Government." The charges were later dropped, and he became known as the pioneer of "New Journalism."

After his conviction, Rusty became enamored with Penny Vick whose daddy owned V. V. Vick Jewelers in downtown Columbus. My sister-in-law Sue said, "The funniest thing was that momma would not let Rusty and Penny sleep together at the cabin until they got married.

It was a running joke until the wedding." On May 16, 1976, the New York Times reported Rusty and Penny were married in Columbus "in a candlelight ceremony at St. Paul United Methodist Church. Mr. Calley, who is 32 years old, met his 26-year-old bride several years ago when he was a customer in one of her father's stores. She is a buyer for her father. It was the first marriage for both." Calley became the gemologist and the manager of their company store in the Cross Country Plaza Mall in Columbus. In July of '83, the Times reported that Calley lived near Ft. Benning and worked six days a week for the store. He was its most popular salesman, laughing and joking, but was reclusive to the public off the job. He was often seen driving around town and working in the yard of the three-bedroom "bungalow" which they bought in '77, but he would not talk to reporters.

Penny and Rusty were divorced in 2006, and he moved to Atlanta to live with their son. My sister-in-law said, "Last we heard from him, Rusty was sick, and we think he died. We haven't heard anything else about him."

What did Sue and her family think of him? Was he a monster? Sue said, "Oh, no. He was the sweetest thing in the world. Even our children loved him; they called him Cousin Rusty. But then, sometimes he would get pretty excited about something. You know how military people are. Once they get that first chance to be in charge, they think they should always be in charge." She laughed, "Like you."

The best friend Rusty Calley probably ever had invited him to speak to the Kiwanis Club of Greater Columbus. So, on August 20, 2009, Rusty dressed up and went to the club's weekly meeting. To prevent a media circus, the press and the public had not been invited. Rusty told the club with tears in his eyes that "there is not a day that goes by that I do not feel remorse for what happened that day in My Lai. I feel remorse for the Vietnamese who were killed, for their families, for the American soldiers involved and for their families. I am very sorry." The place was silent; no one asked any questions. The club members appreciated his coming to speak and let him off easy. They knew it was hard for him, but it didn't change anything; it didn't mean he was innocent.

I think Calley was sincere in his sorrow, but I could be wrong. Time magazine correspondent Peter Range spent many hours with Calley during court breaks and in his apartment and agrees, "That Calley is remorseful is beyond question. He stands convicted of wanton slaughter, but in his private life he is an unlikely villain. Calley has no history of anything more violent than water-skiing." Could only a psychopath or a sociopath not be sorry for doing what Calley had done, no matter the reason? Maybe he was a sociopath, a hot-headed person who lacks remorse, shame or guilt and acts without thinking or caring how anyone else is affected by his actions. Or maybe he was a psychopath, a cold-hearted, calculating person who plots his moves and uses aggression to achieve his goals. Could it be possible that Lt. Calley craved the approval of Captain Medina so much that he was willing to kill innocent women and children to gain it? Medina, the man Calley admired even though he often called him "Sweetheart" in front of his men, may have been as guilty and as much a psychopath as Calley may have been a sociopath. In any event, it's obvious that Rusty Calley was just an average, run-of-the-mill 25-year-old college drop-out who should never have been wearing the bars of an American Army officer. He was untrained, unprepared and incompetent for the onerous task of leading men in the life and death struggle of battle.

I would hope Rusty Calley was sorry for the pain and embarrassment he brought to all honorable and valiant American servicemen and women who've served our country from its beginning, not just during Vietnam. I hope he was sorry for the pain our servicemen needlessly suffered, and still suffer, by the guilt of association with him and by causing some people to wonder if all servicemen are like him. I hope he was sorry for the sleepless nights some GIs have spent wondering if they had been given similar orders, would they have followed them, even if they had to hold their nose to do so.

When the investigations were finalized, Major General Samuel W. Koster was the highest-ranking U.S. Army officer punished for the My Lai massacre and was demoted to Brigadier General.

Eventually, after 30 years, Warrant Officer Hugh Thompson and his crew members in the 123rd Aviation Battalion of the 23rd Infantry Division were considered heroes and were given the praise, awards, and medals they richly deserved. Thompson suffered PTSD, alcoholism, divorce, and severe nighttime disorder but remained in the Army until '83. Removed from life support after extensive cancer treatments, he died in 2006 at age 62.

PHOTOGRAPHS

The author upon graduating from Great Lakes boot camp in 1967.

The jacket patches of RAG outfit VA-42 and the aircraft carrier USS Constellation (CV-64).

The author renting a car with his buddies to tour Oahu in 1969 while on liberty.

Flight operations in the Gulf of Tonkin in the South China Sea in 1969 aboard the Constellation.

F. Lewis Smith

The author replacing an A6's computer part on the Connie's flight deck in 1969.

A VA-85 A6A returning to the Constellation in 1969.

Neither an Officer nor a Gentleman

The author in his gabardine dress blues standing beside the Connie docked in Japan in 1969.

A view of Hong Kong harbor from the air in 1969.

A street scene in crowded but magnificent downtown Hong Kong in 1969.

Vietnamese refugees trying to enter Hong Kong by sea in 1969.

The author as Sailor of the Month at NAS Oceana between cruises in 1970.

VA-85 and USS Forrestal jacket patches.

The USS Forrestal (CV-59) cruising the Mediterranean Sea in 1971.

The Forrestal at anchor in Barcelona, Spain in 1971. The author's spouse joined him thanks to a Dependent's Flight arranged by the Oceana Wives Club.

A VA-85 A6A being trapped on the Forrestal in the Med.

An F4C Phantom flying beside a Russian Tu-95-14 spy plane in the Med.

Neither an Officer nor a Gentleman

The VA-85 AQ shop in 1971 on the Forrestal.
The author is on the far left posing for pictures for the cruise book.

The Esquire magazine with Lt. Calley's cover given to the author in 1971.

Lt. "Rusty" Calley at the River while on trial at Ft. Benning.

Lt. "Rusty" Calley and his girl Deena at the River during his trial.

Neither an Officer nor a Gentleman

An A6 from squadron VA-196 aboard the Connie in early 1970.

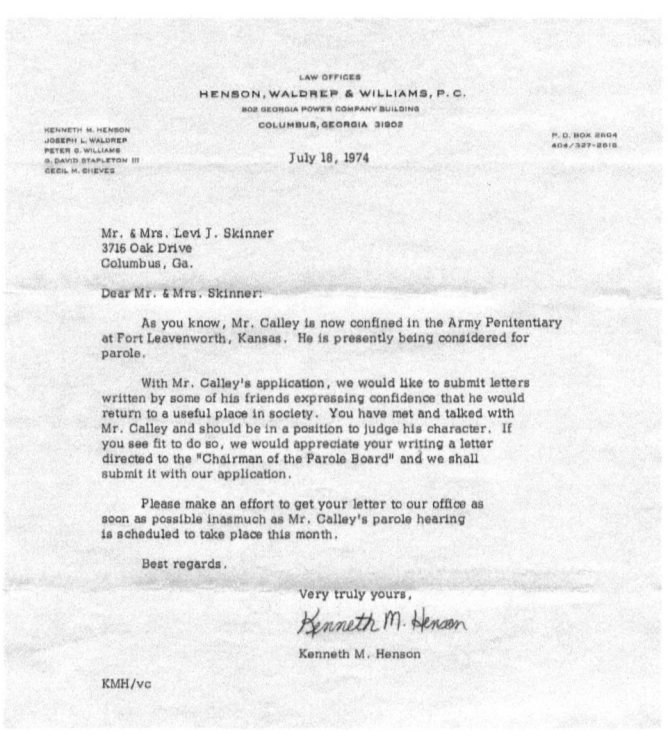

The letter requesting character witnesses for Lt. Calley from his lawyers.

ABOUT THE AUTHOR

F. Lewis Smith is a U.S. Navy Vietnam veteran who served aboard the aircraft carrier USS *Constellation* (CV-64) in 1969-1970 on line in the Gulf of Tonkin. Smith was born and raised in Augusta, Georgia, just two hundred yards down Magnolia Drive from the Augusta National Golf Club's main gate. Smith attended Georgia Tech for two years but, out of money, he joined the Navy for four years in 1967. After Great Lakes, Smith attended Advanced Avionics school at NAS Millington in Memphis and was there when Dr. Martin Luther King, Jr. was assassinated. After training with VA-42 at NAS Oceana, Virginia Beach, VA, Smith was assigned to VA-85 (A6A Black Falcons) for its 1969 Western Pacific deployment.

In 1971, after his USS *Forrestal* (CV-59) carrier duty in the Mediterranean, Smith obtained his Certified Public Accountant license and started his accounting business. His son purchased his practice in 2007, and Smith's passion has been writing historical articles for newspapers and magazines ever since. He published *J. Edgar Thomson: the Georgia Rail Road Years, 1833-1845* in 2017. Smith is the curator of the McDuffie Museum in Thomson, GA, and is on the Board of the Augusta (GA) Jewish Museum.

www.ingramcontent.com/pod-product-compliance
Lightning Source LLC
Chambersburg PA
CBHW030316100526
44592CB00010B/458